T0332154

An Introduction to Nonautonomous Dynamical Systems and their Attractors

INTERDISCIPLINARY MATHEMATICAL SCIENCES*

Published

*For the complete list of titles in this series, please go to
http://www.worldscientific.com/series/ims

Interdisciplinary Mathematical Sciences – Vol. 21

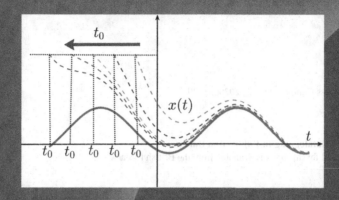

An Introduction to Nonautonomous Dynamical Systems and their Attractors

Peter E Kloeden

Universität Tübingen, Germany

Meihua Yang

Huazhong University of Science and Technology, China

 **World Scientific**

NEW JERSEY · LONDON · SINGAPORE · BEIJING · SHANGHAI · HONG KONG · TAIPEI · CHENNAI · TOKYO

Published by

World Scientific Publishing Co. Pte. Ltd.

5 Toh Tuck Link, Singapore 596224

USA office: 27 Warren Street, Suite 401-402, Hackensack, NJ 07601

UK office: 57 Shelton Street, Covent Garden, London WC2H 9HE

Library of Congress Control Number: 2020051829

British Library Cataloguing-in-Publication Data
A catalogue record for this book is available from the British Library.

Interdisciplinary Mathematical Sciences — Vol. 21
AN INTRODUCTION TO NONAUTONOMOUS DYNAMICAL SYSTEMS AND THEIR ATTRACTORS

ISBN 978-981-122-865-0 (hardcover)
ISBN 978-981-122-866-7 (ebook for institutions)
ISBN 978-981-122-867-4 (ebook for individuals)

For any available supplementary material, please visit
https://www.worldscienti ic.com/worldscibooks/10.1142/12053#t=suppl

Printed in Singapore

Dedicated to the memory of Karin Wahl-Kloeden   (PEK)

Dedicated to my parents and my daughter   (MHY)

Preface

The nature of time in nonautonomous dynamical systems is very different from that in autonomous systems, which depends only on the time that has elapsed since starting rather than on the actual time itself. Consequently, limiting objects may not exist in actual time as in autonomous systems. New concepts of attractors in nonautonomous dynamical systems are thus required.

In addition, the definition of a dynamical system itself needs to be generalised to the nonautonomous context. Here two possibilities are considered: two-parameter semigroups or processes and the skew product flows. Their attractors are defined in terms of families of sets that are mapped onto each other under the dynamics rather than a single set as in autonomous systems. Two types of attraction are now possible: pullback attraction, which depends on the behaviour from the system in the distant past, and forward attraction, which depends on the behaviour of the system in the distant future. Pullback and forward attractors are invariant families of sets which depend on time and exist in actual time for all time. They are generally independent of each other.

The theory of pullback attractors is now very well developed. In contrast the theory of forward attractors is not as satisfactory, since forward attractors often do not exist and when they do, they are often not unique. Moreover, they require the dynamical system to be defined for all time, including the distant past, even though they are concerned with the distant future. The asymptotic behaviour in the future limit seems to be better characterised by omega-limit sets, in terms of which form what are called forward attracting sets. They are generally not invariant in the conventional sense, but are asymptotically positively invariant in general and, if the future dynamics is appropriately uniform, also asymptotically negatively invariant.

Much of this book is based on lectures given by the authors in Frankfurt and Wuhan. It was written mainly when the first author held a "Thousand Expert" Professorship at the Huazhong University of Science and Technology in Wuhan. The financial support from this program and the NSFC grants 11571125 and 11971184 is gratefully acknowledged.

The material in the first nine chapters is now quite well known in the literature, see for example the monograph *Nonautonomous Dynamical Systems* [Kloeden and Rasmussen (2011)]. The material here is restricted to attractors. The emphasis is on systems with compact absorbing sets, thus essentially finite dimensional, and the presentation is also more didactical. This makes the book suitable for mathematics students at upper bachelors or master level and, in particular, for readers from other disciplines, who have a basic background in mathematics.

Chapters 10–12 and Chapter 14 are based on more recent research results of the authors and their coworkers.

Apart from some key references, most of the relevant literature is cited and discussed in the Notes at the end of the book.

The authors thank Hongyong Cui, Christian Pötzsche and Larissa Serdukova for many useful suggestions and assistance. They also thank the referees for their critical reading of the manuscript and insightful comments.

Tübingen   Peter Kloeden
Wuhan   Meihua Yang
August 2020

Peter E. Kloeden
Mathematisches Institut
Universität Tübingen
D-72076 Tübingen
Germany

Meihua Yang
School of Mathematics and Statistics
Huazhong University of Science and Technology
Wuhan 430074
China

Contents

PART 1
Dynamical systems

Chapter 1

Autonomous dynamical systems

Autonomous dynamical systems are briefly recalled here for use, comparison and contrast when we generalise them to the nonautonomous case.

1.1 Autonomous difference equations

Difference equations are a special kind of recursion relation. The adjective *difference* has historical origins from when such equations were expressed in terms of the difference operator $\Delta x_n = x_{n+1} - x_n$ in contrast to the differential operator in differential equations.

Let $f : \mathbb{R}^d \to \mathbb{R}^d$. Then

$$x_{n+1} = f(x_n), \qquad n = 0, 1, 2, \ldots, \tag{1.1}$$

is a first-order *autonomous difference equation* on $\mathbb{R}^d$. It is called autonomous because f does not depend explicitly on n. The mapping f will be assume to be continuous here.

A simple example is the scalar population growth equation

$$x_{n+1} = a x_n (1 - x_n), \qquad n = 0, 1, 2, \ldots,$$

where a is a positive constant and $f(x) = ax(1 - x)$ with $x \in \mathbb{R}^1$.

The solution of the difference equation (1.1) is given through succesive iteration by

$$x_n = f^n(x_0) := \underbrace{f \circ f \circ \cdots \circ f}_{n \text{ times}}(x_0), \qquad n = 0, 1, 2, \ldots,$$

with f^0 defined to be the identity mapping, i.e., $f^0(x) \equiv x$.

Define the *solution mapping* $\pi : \mathbb{Z}^+ \times \mathbb{R}^d \to \mathbb{R}^d$ by

$$\pi(n, x_0) = f^n(x_0).$$

This mapping has the properties

i) *initial condition* $\pi(0, x_0) = x_0$ for all $x_0 \in \mathbb{R}^d$,

ii) *semi-group under composition:*
$\pi(n + m, x_0) = \pi(n, \pi(m, x_0))$ for all $m, n \in \mathbb{Z}^+$ and $x_0 \in \mathbb{R}^d$,

iii) *continuity:* the mapping $x_0 \mapsto \pi(n, x_0)$ is continuous for each $n \in \mathbb{Z}^+$.

The initial condition is obvious, while continuity follows by the fact that the composition of continuous mappings is continuous.

To verify the semi-group property note that

$$\pi(n+m, x_0) = f^{n+m}(x_0) := \underbrace{f \circ f \circ \cdots \circ f}_{n+m \text{ times}}(x_0)$$

$$= \underbrace{f \circ f \circ \cdots \circ f}_{n \text{ times}} \circ \underbrace{f \circ f \circ \cdots \circ f}_{m \text{ times}}(x_0)$$

$$= f^n \circ f^m(x_0) = \pi(n, \pi(m, x_0))$$

for all m, $n \in \mathbb{Z}^+$ and $x_0 \in \mathbb{R}^d$, by the associativity of the composition operation.

Remark 1.1. The mapping f need not be invertible, e.g., $f(x) = x(2-x)$ on $\mathbb{R}$.

Suppose that f is a *homeomorphism*, i.e., f is continuous and its inverse f^{-1} exists and is continuous. Then we can define $\pi(n, x_0)$ for $n = -1, -2, -3, \ldots$ too, i.e., we can extend π from $\mathbb{Z}^+$ to $\mathbb{Z}$: for $n < 0$ define

$$\pi(n, x_0) = \underbrace{f^{-1} \circ f^{-1} \circ \cdots \circ f^{-1}}_{-n \text{ times}}(x_0) = \left(f^{-1}\right)^{-n}(x_0),$$

i.e., for $n < 0$ we interpret f^n as $\left(f^{-1}\right)^{-n}$.

Then π satisfies the *group property* under composition. The main new thing to check is the case $n < 0 < m$.

$$\pi(n, \pi(m, x_0)) = \left(f^{-1}\right)^{-n} \circ f^m(x_0) = \underbrace{f^{-1} \circ \cdots \circ f^{-1}}_{-n \text{ times}} \circ \underbrace{f \circ f \circ \cdots \circ f}_{m \text{ times}}(x_0),$$

$$= \begin{cases} \left(f^{-1}\right)^{-n-m}(x_0) & \text{if } n+m < 0, \\ f^{n+m}(x_0) & \text{if } n+m > 0 \end{cases}$$

$$= \pi(n+m, x_0)).$$

Remark 1.2. The mapping f given by $f(x) = x(2-x)$ maps the unit interval $[0, 1]$ into itself. It is invertible on this interval and generates a group under composition on it.

1.2   Discrete-time autonomous dynamical systems

More generally, instead of $\mathbb{R}^d$, we could consider a *metric space* (X, d_X) as the state space.

Definition 1.1. A discrete-time autonomous dynamical system on a state space X is given by mapping $\pi : \mathbb{Z} \times X \to X$, which satisfies the properties:

i) *initial condition:* $\pi(0, x_0) = x_0$ for all $x_0 \in X$,

ii) *group under composition:*
$$\pi(n + m, x_0) = \pi(n, \pi(m, x_0)) \text{ for all } m, n \in \mathbb{Z} \text{ and } x_0 \in X,$$

iii) *continuity:* the mapping $x_0 \mapsto \pi(n, x_0)$ is continuous for each $n \in \mathbb{Z}$.

We have a *semi-dynamical system* if $\mathbb{Z}$ is replaced by $\mathbb{Z}^+ \cup \{0\}$ and ii) is replaced by the *semi-group property*, i.e., it holds just for n and $m \in \mathbb{Z}^+ \cup \{0\}$.

Note that the family of mappings $\{\pi(n, \cdot)\}_{n \in \mathbb{Z}}$ is a group of mappings from X into itself under composition in the first case, while $\{\pi(n, \cdot)\}_{n \in \mathbb{Z}^+ \cup \{0\}}$ is a semi-group of mappings under composition.

The following example of a discrete-time dynamical system will be important later in formulating nonautonomous difference equations generated by a finite number of mappings as a skew product flow, see Subsection 3.3.3.

Example 1.1. Let $X := \{1, N\}^{\mathbb{Z}}$, where $N \in \mathbb{N}$, be the space of bi-infinite integer-valued sequences $\mathbf{x} = (\ldots, x_{-1}, x_0, x_1, \ldots)$ with $x_j \in \{1, \ldots, N\}$. Then X forms a compact metric space with the metric (the proof is left as an exercise)

$$d_X(\mathbf{x}, \mathbf{y}) := \sum_{n \in \mathbb{Z}} 2^{-|n|} |x_n - y_n|.$$

Let $\theta : X \to X$ be the *left shift operator*, which is defined as

$$\widetilde{\mathbf{x}} := \theta(\mathbf{x}), \qquad \widetilde{x}_n := (\theta(\mathbf{x}))_n = x_{n+1}, \quad n \in \mathbb{Z}^+ \cup \{0\},$$

$$\mathbf{x} = (\ldots, x_{-2}, x_{-1}, x_0, x_1, x_2, \ldots)$$

$$\downarrow$$

$$\widetilde{\mathbf{x}} = \underbrace{(\ldots, x_{-1}, x_0, x_1, x_2, x_3, \ldots)}.$$
$$\leftarrow \text{shifted to the left}$$

Note that θ is continuous and has a continuous inverse θ^{-1} given by the *right shift operator*

$$\left(\theta^{-1}(\mathbf{x})\right)_n := (\theta(\mathbf{x}))_{n-1}, \quad n \in \mathbb{Z}^+ \cup \{0\}.$$

Then $\pi : \mathbb{Z} \times X \to X$ defined by

$$\pi(n, \mathbf{x}) = \begin{cases} \left(\theta^{-1}\right)^{-n}(\mathbf{x}) & \text{if } n < 0, \\ \theta^n(\mathbf{x}) & \text{if } n \geq 0 \end{cases}$$

is a discrete-time autonomous dynamical system on X. In particular, the family $\{\theta^n : n \in \mathbb{Z}\}$ is a group under composition.

This example can be represented as a first-order difference equation on the sequence space X with the mapping $f : X \to X$ defined by $f(\mathbf{x}) = \pi(1, \mathbf{x})$.

1.3   Continuous-time autonomous dynamical systems

Dynamical systems defined on the time set $\mathbb{R}$ instead of $\mathbb{Z}$ are called continuous-time dynamical system.

Consider the autonomous ordinary differential equation (ODE) on $\mathbb{R}^d$

$$\frac{dx}{dt} = f(x), \qquad x \in \mathbb{R}^d, \tag{1.2}$$

and assume that the solution $x(t) = x(t, t_0, x_0)$ starting at $x(t_0) = x_0$ exists for all $x_0 \in \mathbb{R}^d$ and $t, t_0 \in \mathbb{R}$, is unique, and continuous in the initial value (t_0, x_0). (This is ensured if f satisfies a global Lipschitz condition or, more generally, satisfies a local Lipschitz condition and some kind of growth condition to prevent solutions from exploding in finite time.)

Note that $x(t; t_0, x_0) \equiv x(t - t_0; 0, x_0)$, i.e., the solutions of an autonomous ODE depend only on the *elapsed time* $t - t_0$ and not on the actual times t and t_0 separately. This is a characteristic property of autonomous systems.

Example 1.2. The scalar ODE

$$\frac{dx}{dt} = x$$

with initial value $x(t_0) = x_0$ has the explicit solution

$$x(t, t_0, x_0) = x_0 e^{t-t_0}.$$

This means that the solution curves are invariant under time translation.

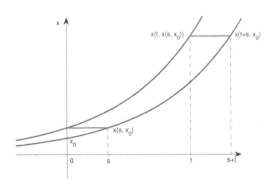

Fig. 1.1   Invariance of solution curves under time translation.

Henceforth we will write just $x(t - t_0, x_0)$ for $x(t, t_0, x_0)$ and consider the solution mapping x as a mapping from $\mathbb{R} \times \mathbb{R}^d$ into $\mathbb{R}^d$. This mapping has the following important properties:

1)   *initial condition* $x(0, x_0) = x_0$.

2) *group under composition* $x(t + s, x_0) = x(t, x(s, x_0))$ for all t, $s \in \mathbb{R}$.

These correspond to the uniqueness of solutions to initial values problems.

3) *continuity* $(t, x_0) \mapsto x(t, x_0)$ is continuous.

Remark 1.3. The solution of the ODE is in fact differentiable in t, hence continuous. From the theory of differential equations we know that it is also continuous in the initial condition x_0.

1.4 General case

Again, instead of $\mathbb{R}^d$, we consider a general *complete metric space* (X, d_X) as the state space.

In addition, we consider the time set $\mathbb{T}$, which is $\mathbb{Z}$ in the discrete-time case and $\mathbb{R}$ in the continuous-time case. Similarly, $\mathbb{T}^+$ denotes $\mathbb{Z}^+$ and $\mathbb{R}^+$ in these cases.

Definition 1.2. An autonomous dynamical system on a state space X is given by mapping $\pi : \mathbb{T} \times X \to X$, which satisfies the properties:

i) *initial condition:* $\pi(0, x_0) = x_0$ for all $x_0 \in X$,
ii) *group under composition:*
   $\pi(s + t, x_0) = \pi(s, \pi(t, x_0))$ for all s, $t \in \mathbb{T}$ and $x_0 \in X$,
iii) *continuity:* the mapping $(t, x) \mapsto \pi(t, x)$ is continuous at all points $(t_0, x_0) \in \mathbb{T} \times X$.

Remark 1.4. If $\mathbb{T} = \mathbb{Z}$, then continuity in t in iii) is with respect to the *discrete topology*, so we essentially just have continuity in x_0 for each t.

For many differential equations the solutions may exist forward in time, but not backwards in time.

Example 1.3. Consider the scalar differential equation

$$\frac{dx}{dt} = x \left(1 - x^2\right)$$

with the explicit solution (see Fig. 1.2)

$$x(t, x_0) = \frac{x_0 e^t}{\sqrt{1 + x_0^2(e^{2t} - 1)}}.$$

The solution mapping $x : \mathbb{R}^+ \times \mathbb{R}^1 \to \mathbb{R}^1$ satisfies the semi-group rather than group property since the solutions do not exist for all negative time when $|x_0| > 1$.

A similar situation arises if the mapping f in the difference equation (1.1) is not invertible.

These examples motivate the definition of an abstract *semi-dynamical system* π on the state space X, in which the time set $\mathbb{T}$ is replaced by $\mathbb{T}^+ \cup \{0\}$ and the group property (ii) is replaced by the *semi-group property* over $\mathbb{T}^+ \cup \{0\}$.

Definition 1.3. An autonomous semi-dynamical dynamical system on a state space X is given by mapping $\pi : \mathbb{T}^+ \cup \{0\} \times X \to X$, which satisfies the properties:

i) *initial condition:* $\pi(0, x_0) = x_0$ for all $x_0 \in X$,

ii) *semi-group under composition:*
$\pi(s + t, x_0) = \pi(s, \pi(t, x_0))$ for all $s, t \in \mathbb{T}^+ \cup \{0\}$ and $x_0 \in X$,

iii) *continuity:* the mapping $(t, x) \mapsto \pi(t, x)$ is continuous at all points $(t_0, x_0) \in \mathbb{T}^+ \cup \{0\} \times X$.

1.5   Invariant sets, omega limit sets and attractors

We are interested in the *long term*, i.e., asymptotic, behaviour of the system, so we will consider a semi-dynamical system π on a complete metric state space (X, d_X).

The *invariant sets* in X w.r.t. π provide us with a lot of useful information about the dynamical behaviour of π.

Definition 1.4. A nonempty subset A of X is called an *invariant set* of a semi-dynamical system π if

$$\pi(t, A) = A \quad \text{for all } t \in \mathbb{T}^+,$$

where

$$\pi(t, A) := \bigcup_{a \in A} \{\pi(t, a)\}.$$

Example 1.4. Consider the scalar ODE

$$\frac{dx}{dt} = x \left(1 - x^2\right), \tag{1.3}$$

which generates semi-dynamical system on $X = \mathbb{R}$. It has steady state (i.e., constant) solutions $\bar{x} = 0, \pm 1$.

These steady states are invariant sets as are combinations of them: $\{0\}$, $\{-1\}$, $\{1\}$, $\{-1, 1\}$, $\{0, 1\}$, $\{0, -1\}$, $\{0, -1, 1\}$. Other closed invariant sets are the intervals $[-1, 0]$, $[0, 1]$, $[-1, 1]$.

One could also consider unions of intervals and disjoint steady states such as $\{-1\} \cup [0, 1]$. In addition, one has invariant sets such as $(-\infty, -1]$ and $[1, \infty)$.

The set $[-1, 1]$ is the largest compact invariant set for this example. It has a special significance.

1.5.1   *Omega-limit sets*

The ω-limit sets of a semi-dynamical system characterise its asymptotic behaviour as $t \to \infty$.

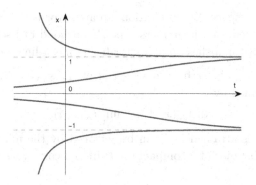

Fig. 1.2   Steady states and typical solutions of the ODE (1.3).

Definition 1.5. (Omega-limit sets) The ω-*limit set* of a bounded set $B \subset X$ is defined by

$$\omega(B) = \{x \in X \ : \ \exists t_k \to \infty, y_k \in B \ \text{ with } \ \pi(t_k, y_k) \to x\}.$$

When $B = \{y\}$, i.e., just one point, one usually writes $\omega(y)$ instead of $\omega(\{y\})$. In general,

$$\omega(B) \supsetneq \bigcup_{y \in B} \omega(y).$$

The ω-limit sets have the following properties.

Theorem 1.1. *For a nonempty bounded subset B of X,*

$$\omega(B) = \bigcap_{t \geq 0} \overline{\bigcup_{s \geq t} \pi(s, B)} \ .$$

The next theorem is stated for the state space $X = \mathbb{R}^d$. A similar result holds in more general state spaces provide π satisfies an appropriate compactness property, see Remark 1.5.

Theorem 1.2. *Let π be a semi-dynamical system on $\mathbb{R}^d$. If $\bigcup_{t \geq 0} \pi(t, B)$ is bounded, then the ω-limit set $\omega(B)$ for every nonempty bounded set $B \subset X$ is nonempty, compact and invariant.*

Note in Theorem 1.2 that $\omega(B)$ is connected for continuous time systems when B is connected. It need not, however, be connected for discrete time systems, e.g., consider the autonomous difference equation with $f(x) = -x$, so $f^n(-1) = (-1)^{n+1}$.

1.5.2   *Attractors*

An attractor is an invariant set of special interest since contains all the long term dynamics of a dissipative dynamical system, i.e., it is where every thing ends up. In particular, it contains the omega limit set $\omega(D)$ of every nonempty bounded subset D of the state space X. In addition, an attractor is the omega limit set of a neighbourhood of itself, i.e., it attracts a neighbourhood of itself. This additional stability property distinguishes an attractor from omega limit sets in general.

The distance of a point x from a subset A of X, where (X, d_X) is a complete metric space, is defined by

$$\operatorname{dist}_X(x, A) := \inf_{a \in A} d_X(x, a),$$

where the infimum is attained and can be replaced by the minimum when A is a compact set. The distance of a compact set B from a compact set A is then defined by

$$\operatorname{dist}_X(B, A) := \max_{b \in B} \operatorname{dist}_X(b, A) = \max_{b \in B} \inf_{a \in A} d_X(b, a).$$

In general, $\operatorname{dist}_X(A, B) \neq \operatorname{dist}_X(B, A)$, for example,

$$0 = \operatorname{dist}_{\mathbb{R}^1}(\{0\}, [0, 1]) \neq \operatorname{dist}_{\mathbb{R}^1}([0, 1], \{0\}) = 1$$

for the subsets $A = \{0\}$ and $B = [0, 1]$ of $\mathbb{R}^1$.

Definition 1.6. A (global) *attractor* of a semi-dynamical system π is nonempty compact invariant set A of X which attracts all nonempty bounded subsets D of X, i.e.,

$$\operatorname{dist}_X(\pi(t, D), A) \to 0 \quad \text{as } t \to \infty.$$

The global attractor in Example 1.4 is $A = [-1, 1]$. It is the largest compact invariant set that attracts all other nonempty bounded subsets. The set $\{1\}$ is a *local* attractor in this example since it attracts only bounded subsets of the positive half line. We will mainly talk about global attractors and often omit the adjective global.

An attractor may have a very complicated geometrical shape, e.g., the fractal dimensional set in the Lorenz ODE system. It is often easier to determine an absorbing set with a simpler geometrical shape such as a ball.

Definition 1.7. A nonempty subset B of X is called an *absorbing set* of π if for every nonempty bounded subset D of X there exists a $T_D \geq 0$ such that

$$\pi(t, D) \subset B$$

for all $t \geq T_D$.

All of the future dynamics is in B, which need not be invariant, but often it is *positively invariant*, i.e., $\pi(t, B) \subset B$ for all $t \in \mathbb{T}^+$.

Theorem 1.3. *If a semi-dynamical system π on a complete metric space (X, d_X) has a compact absorbing set $B \subset X$, then it has an attractor A, which is contained in B and is given by*

$$A = \bigcap_{t \geq 0} \overline{\bigcup_{s \geq t} \pi(s, B)}.$$

If B is positively invariant then

$$A = \bigcap_{t \geq 0} \pi(t, B).$$

The proof is a special case on the corresponding proof for nonautonomous dynamical systems that will be given later.

Note that the sets

$$A_t := \overline{\bigcup_{s \geq t} \pi(s, B)}$$

are *compact* as closed subsets of the compact set B (provided t is large enough) and *nested*, i.e., $A_s \subset A_t$ if $s \geq t$. This means that A exists and is a nonempty compact subset of B. We will say more about attractors of autonomous dynamical systems in Chapter 5

Remark 1.5. Theorem 1.3 is useful in locally compact state spaces such as $\mathbb{R}^d$ where the closed and bounded subsets are the compact subsets. In infinite dimensional dynamical systems the absorbing sets are usually assumed to be just closed and bounded, which are more common and easily determined than compact sets in such spaces. Some additional compactness property of the semi-group π such as its asymptotic compactness is then needed to ensure the nonemptiness of the attractor in Theorem 1.3.

Definition 1.8. A semi-dynamical system π on a complete metric space (X, d_X) is said to be *asymptotically compact* if, for every sequence $\{s_k\}_{k \in \mathbb{N}}$ in $\mathbb{R}^+$ with $s_k \to \infty$ as $k \to \infty$ and every bounded sequence $\{x_k\}_{k \in \mathbb{N}}$ in X, the sequence $\{\pi(s_k, x_k)\}_{k \in \mathbb{N}}$ has a convergent subsequence.

Chapter 2

Nonautonomous dynamical systems: processes

The properties of the solution mappings of nonautonomous difference and differential equations are considered here to motivate the formulation of a nonautonomous dynamical system as a process of 2-parameter semi-group.

2.1 Nonautonomous difference equations

A nonautonomous difference equation on $\mathbb{R}^d$ has the form

$$x_{n+1} = f_n(x_n),$$

where the $f_n : \mathbb{R}^d \to \mathbb{R}^d$ are continuous mappings.

The mappings f_n may be completely different from each other, but in concrete applications they are often similar but with different parameters. For example, let $d = 1$ and $f_n(x) = a_n x(1-x)$, where $a_n \in \mathbb{R}^+$. Here

$$x_{n+1} = a_n x_n(1 - x_n),$$

is a population growth model with a (possibly) different growth rate a_n each year.

As another example, the Euler scheme with different step sizes $\Delta_n > 0$ for an autonomous ODE

$$\frac{dx}{dt} = f(x), \qquad x \in \mathbb{R}^d, \tag{2.1}$$

is given by

$$x_{n+1} = x_n + f(x_n)\Delta_n.$$

Here $f_n(x) = x + f(x)\Delta_n$ with the parameter Δ_n.

In general, the mappings f_n need not be invertible.

The solution mapping

Consider a nonautonomous difference equation

$$x_{n+1} = f_n(x_n)$$

on $\mathbb{R}^d$ with the initial value $x_{n_0} = x_0$. Then $x_{n_0+1} = f_{n_0}(x_{n_0})$ and

$$x_{n_0+2} = f_{n_0+1}(x_{n_0+1}) = f_{n_0+1}(f_{n_0}(x_{n_0})) = \underbrace{f_{n_0+1} \circ f_{n_0}(x_{n_0})}_{\text{composition}}.$$

Similarly,

$$x_n = \underbrace{f_{n-1} \circ \cdots \circ f_{n_0}(x_{n_0})}_{\text{composition}}$$

for all $n \geq n_0 + 1$. This defines a *solution mapping*

$$\phi(n, n_0, x_0) := f_{n-1} \circ \cdots \circ f_{n_0}(x_{n_0})$$

for all $n \geq n_0 + 1$ in $\mathbb{Z}$ and $x_0 \in \mathbb{R}^d$ with $\phi(n_0, n_0, x_0) := x_0$.

The solution mappping $\phi(n, n_0, x_0)$ has the following properties:

1) *initial condition* $\phi(n_0, n_0, x_0) = x_0$. (This is part of the definition above.)

2) *2-parameter semi-group under composition*

$$\phi(n_2, n_0, x_0) = \phi(n_2, n_1, \phi(n_1, n_0, x_0))$$

for all $n_0 \leq n_1 \leq n_2$ in $\mathbb{Z}$ and $x_0 \in \mathbb{R}^d$.

This follows from the *associativity of composition*

$$\phi(n_2, n_0, x_0) = f_{n_2-1} \circ \cdots \circ f_{n_0}(x_{n_0})$$

$$= f_{n_2-1} \circ \cdots \circ f_{n_1}(f_{n_1-1} \circ \cdots \circ f_{n_0}(x_{n_0}))$$

$$= \phi(n_2, n_1, \underbrace{f_{n_1-1} \circ \cdots \circ f_{n_0}(x_{n_0})}_{=x_{n_1}}).$$

3) *continuity* $x_0 \mapsto \phi(n_1, n_0, x_0)$ is continuous for all $n_1 \geq n_0$.

This follows from the continuity of the composition of continuous functions

$$x \mapsto f_{n_1-1} \circ \cdots \circ f_{n_0}(x), \quad n_1 > n_0.$$

2.2 Nonautonomous differential equations

Consider a *nonautonomous ODE* on $\mathbb{R}^d$

$$\frac{dx}{dt} = f(t, x), \qquad t \in \mathbb{R}, \ x \in \mathbb{R}^d,$$

with initial value $x(t_0) = x_0$. Assume the existence and uniqueness of the solutions and their continuity in initial conditions. In particular, the solution mapping $x(t) = x(t, t_0, x_0)$ is then defined for all $x_0 \in \mathbb{R}^d$ and $t \geq t_0$ in $\mathbb{R}$. (This is guaranteed by appropriate assumptions on the vector field f, e.g., if it is continuous in both variables and satisfies a global Lipschitz condition in the second variable uniformly in the first, though weaker conditions are possible.)

Example 2.1. The nonautonomous ODE on $\mathbb{R}$,

$$\frac{dx}{dt} = 2tx$$

has the solution $x(t, t_0, x_0) = x_0 e^{t^2 - t_0^2}$.

Unlike as an autonomous ODE, the solution here depends explicitly on both t and t_0, and not just on the elapsed time since starting $t - t_0$. In particular, here $t^2 - t_0^2 = (t - t_0)(t + t_0)$ cannot be written in terms of $t - t_0$ only.

The solution mappping $x(t, t_0, x_0)$ has the following properties:

1) *initial condition:* $x(t_0, t_0, x_0) = x_0$.

2) *2-parameter semi-group under composition:*

$$x(t_2, t_0, x_0) = x(t_2; t_1, x(t_1, t_0, x_0))$$

for all $t_0 \leq t_1 \leq t_2$ in $\mathbb{R}$.

This is a consequence of the existence and uniqueness of solutions. The solution starting at (t_1, x_1), where $x_1 = x(t_1, t_0, x_0)$ is unique, so we have

$$x(t_2, t_0, x_0) = x(t_2, t_1, x_1) = x(t_2, t_1, x(t_1, t_0, x_0)),$$

This is called the *2-parameter semi-group property* , i.e., with parameters t and t_0 instead of just the elapsed time $t - t_0$ in an autonomous system.

3) *continuity:* $(t, t_0, x_0) \mapsto x(t, t_0, x_0)$ is continuous.

The solution of the ODE is differentiable in t, hence continuous. From the theory of differential equations we know that it is also continuous in the initial condition (t_0, x_0).

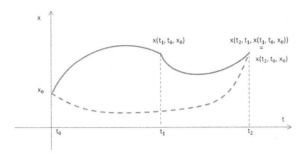

Fig. 2.1   Under the 2-parameter semi-group property the dashed full solution and the concatenated solutions coincide.

2.3   Abstract nonautonomous dynamical systems

We will consider two abstract formulations of nonautonomous dynamical systems in this book. The first is a more direct generalisation of the definition of an abstract autonomous semi-dynamical system and is based on the properties of the solution mappings of nonautonomous differential and difference equations. It is called a *process* or *2-parameter semi-group*. The other case, called a *skew product flow*, will be introduced in Chapter 9.

Instead of $\mathbb{R}^d$ we consider a *complete metric space* (X, d_X) as the state space. In addition, we consider a time set $\mathbb{T}$, which is $\mathbb{Z}$ in the discrete-time case and $\mathbb{R}$ in the continuous-time case. Similarly, $\mathbb{T}^+$, denotes $\mathbb{Z}^+$ and $\mathbb{R}^+$ in these time cases. Define

$$\mathbb{T}^2_{\geq} := \{(t, t_0) \in \mathbb{T} \times \mathbb{T} : t \geq t_0\}.$$

Definition 2.1. (Process) A *process* is a mapping ϕ from $\mathbb{T}^2_{\geq} \times X \to X$ with the following properties:

i) *initial condition:* $\phi(t_0, t_0, x_0) = x_0$ for all $x_0 \in X$ and $t_0 \in \mathbb{T}$.
ii) *2-parameter semi-group property:*

$$\phi(t_2, t_0, x_0) = \phi(t_2, t_1, \phi(t_1, t_0, x_0))$$

for all $(t_1, t_0), (t_2, t_1) \in \mathbb{T}^2_{\geq}$ and $x_0 \in X$.
iii) *continuity:* the mapping $(t, t_0, x_0) \mapsto \phi(t, t_0, x_0)$ is continuous.

The continuity $(t, t_0, x_0) \mapsto \phi(t, t_0, x_0)$ in the discrete-time case $\mathbb{T} = \mathbb{Z}$ reduces to continuity $x_0 \mapsto \phi(t, t_0, x_0)$ in x_0 for all $t \geq t_0$ in $\mathbb{T}$.

Remark 2.1. We can consider a process ϕ as a 2-parameter family of mappings $\phi_{t,t_0}(\cdot)$ on X that forms a *2-parameter semi-group* under composition, i.e.,

$$\phi_{t_2,t_0}(x) = \phi_{t_2,t_1} \circ \phi_{t_1,t_0}(x)$$

for all $t_0 \leq t_1 \leq t_2$ in $\mathbb{T}$.

Remark 2.2. For an autonomous system, a process reduces to

$$\pi(t - t_0, x_0) = \phi(t, t_0, x_0),$$

since the solutions depend only on the elapsed time $t - t_0$, i.e., just one parameter instead of independently on the actual time t and the initial time t_0, i.e., two parameters.

2.4   A process as an autonomous semi-dynamical system

We can consider a process ϕ on a state space X and time set $\mathbb{T}$ as an autonomous semi-dynamical system on the extended state space $\mathfrak{X} := \mathbb{T} \times X$.

Define $\Pi : \mathbb{T}^+ \times \mathfrak{X} \to \mathfrak{X}$ by

$$\Pi(\tau, (t_0, x_0)) := (\tau + t_0, \phi(\tau + t_0, t_0, x_0))$$

for all $t \in \mathbb{T}$ and $(t_0, x_0) \in \mathfrak{X}$.

i) *initial condition:*   For $\tau = 0$ we have

$$\Pi(0, (t_0, x_0)) := (t_0, \phi(t_0, t_0, x_0)) = (t_0, x_0),$$

which is just the initial condition property.

ii) *semi-group property:*   For $\sigma, \tau \in \mathbb{T}^+$, using the 2-parameter semi-group property in the second to third line, we have

$$\Pi(\sigma + \tau, (t_0, x_0)) = (\sigma + \tau + t_0, \phi(\sigma + \tau + t_0, t_0, x_0))$$

$$= (\sigma + (\tau + t_0), \phi(\sigma + (\tau + t_0), \tau + t_0, \phi(\tau + t_0, t_0, x_0))$$

$$= \Pi(\sigma, (\tau + t_0, \phi(\tau + t_0, t_0, x_0)) = \Pi(\sigma, \Pi(\tau, (t_0, x_0))).$$

This is the semi-group property for Π on the extended state space $\mathfrak{X}$.

iii) *continuity :*   The continuity of Π follows directly from that of the constituent functions, i.e.,

$$(\tau, (t_0, x_0)) \longrightarrow \Pi(\tau, (t_0, x_0)) := (\tau + t_0, \phi(\tau + t_0, t_0, x_0)).$$

Remark 2.3. This autonomous semi-dynamical system Π on the state space $\mathfrak{X}$ is not very useful in practice since it has no compact invariant sets and hence cannot have an attractor.

To see this note that if $\Pi(t, \mathfrak{A}) = \mathfrak{A}$ for all $t \in \mathbb{T}^+$ then

$$(t + t_0, \phi(t + t_0, t_0, x_0)) = \Pi(t, (t_0, x_0)) \in \mathfrak{A}$$

for all $t \in \mathbb{T}^+$ and $(t_0, x_0) \in \mathfrak{A}$, so $\mathfrak{A}$ is unbounded in its first component and hence is not compact.

Nevertheless, this example will give us insights on how to define invariant sets and attractors for a process. It is a simple (but a not very interesting) example of a skew-product flow, which will be introduced in the next chapter.

Chapter 3

Skew product flows

Skew product flows are a more complicated formalism of nonautonomous dynamical systems than processes, but they have more detailed information built into them and so one can often obtain more specific results. Autonomous triangular systems will first be examined to motivate the definition of a skew product flow.

3.1 Autonomous triangular systems

Consider a system of autonomous ODEs on $\mathbb{R}^m \times \mathbb{R}^d$ with the *triangular structure*

$$\frac{dp}{dt} = g(p), \qquad p \in \mathbb{R}^m, \tag{3.1}$$

$$\frac{dx}{dt} = f(x, p), \qquad x \in \mathbb{R}^d, \, p \in \mathbb{R}^m. \tag{3.2}$$

Assume that forward existence and uniqueness of solutions holds as well as their continuity in initial values. Then the system of ODEs (3.1)–(3.2) generates an autonomous semi-dynamical system

$$\Pi\left(t, (p_0, x_0)\right) = (p(t, p_0), x(t, p_0, x_0))$$

on $\mathbb{R}^m \times \mathbb{R}^d$.

Note that the p-system (3.1) is independent of the x system and generates an autonomous semi-dynamical system on $\mathbb{R}^m$.

We can consider the p-system as the *driving system* in the x-system, which is then *nonautonomous*

$$\frac{dx}{dt} = \tilde{f}_{p_0}(x, t) = f(x, p(t, p_0)) \tag{3.3}$$

for each $p_0 \in \mathbb{R}^m$. We now want to know what the dynamics of this nonautonomous system on $\mathbb{R}^d$ looks like.

Denote the solution mapping of the nonautonomous ODE (3.3) by $\varphi(t, p_0, x_0)$ given the

1) *initial condition:* $\varphi(0, p_0, x_0) = x_0$ for each p_0 and x_0.

We obviously also have

2) *continuity condition:* $(t, p_0, x_0) \mapsto \varphi(t, p_0, x_0)$ is continuous.

What kind of evolution property does φ satisfy? e.g., like a group or 2-parameter semi-group property:

$$\varphi(s + t, p_0, x_0) = ?? \qquad \forall s, t \in \mathbb{R}^+, p_0 \in \mathbb{R}^m, x_0 \in \mathbb{R}^d.$$

Let us look at the corresponding autonomous semi-dynamical system on

$$\Pi\left(t, (p_0, x_0)\right) = (p(t, p_0), x(t, p_0, x_0))$$

on $\mathbb{R}^m \times \mathbb{R}^d$. Its semi-group property reads:

$$\Pi\left(s + t, (p_0, x_0)\right) = \Pi\left(s, \Pi\left(t, (p_0, x_0)\right)\right),$$

so from the definition of Π we have

$$(p(s + t, p_0), x(s + t, p_0, x_0)) = \Pi\left(s + t, (p_0, x_0)\right) = \Pi\left(s, \Pi\left(t, (p_0, x_0)\right)\right)$$

$$= \Pi\left(s, (p(t, p_0), x(t, p_0, x_0))\right)$$

$$= (p(s, p(t, p_0)), x\left(s, p(t, p_0), x(t, x_0, p_0)\right)),$$

i.e., with the new initial condition $(p(t, x_0), x(t, p_0, p_0))$ at time t.

Now the second components are equal by the semi-group property of the driving system, i.e.,

$$p(s + t, p_0) = p(s, p(t, p_0)).$$

Comparing the first components we have

$$x\left(s + t, p_0, x_0\right) = x\left(s, p(t, p_0), x(t, p_0, x_0)\right).$$

The *evolution property* of the solution mapping $\varphi(t, p_0, x_0)$ of the nonautonomous ODE (3.3) thus reads

$$\varphi\left(s + t, p_0, x_0\right) = \varphi\left(s, p(t, p_0), \varphi(t, p_0, x_0)\right),$$

which is called the *cocycle property*.

Remark 3.1. Note that the driving system is updated at time t to $p(t, p_0)$ when the system moves to $\varphi(t, p_0, x_0)$.

Example 3.1. Consider a system of ODEs (3.2)–(3.1) with $m = d = 1$, i.e.,

$$\frac{dp}{dt} = p(1 - p), \qquad \frac{dx}{dt} = -x + p,$$

with the explicit solutions

$$p(t, p_0) = \frac{p_0 e^t}{1 + p_0 \left(e^t - 1\right)}$$

and

$$x(t, x_0, p_0) = x_0 e^{-t} + e^{-t} \int_0^t e^s p(s, p_0) \, ds.$$

Let us restrict the p-system to the invariant set $[0, 1]$. (Note that $\bar{p} = 0$ and 1 are steady state solutions.) Then the p-system is a group, i.e., reversible dynamical system on $[0, 1]$. What happens to $x(t)$?

3.2 Definition of a skew product flow

A skew product flow consists of an *autonomous dynamical system* (full group) on a base space P, which is the source of the nonautonomity in a *cocycle mapping* acting on a state space X. The autonomous dynamical system is often called the *driving system*.

Suppose that (P, d_P) and (X, d_X) are complete metric spaces and consider the time set $\mathbb{T}$.

Definition 3.1. A skew product flow (θ, φ) on $P \times X$ consists of an autonomous dynamical system $\theta = \{\theta_t\}_{t \in \mathbb{T}}$ acting on a complete metric space (P, d_P), which is called the base space or hull, i.e.,

   i) $\theta_0(p) = p$,   *ii)* $\theta_{s+t}(p) = \theta_s \circ \theta_t(p)$,   *iii)* $(t, p) \mapsto \theta_t(p)$ continuous

for all $p \in P$ and $s, t \in \mathbb{T}$, and a cocycle mapping $\varphi : \mathbb{T}^+ \times P \times X \to X$ acting on a complete metric space (X, d_X), which is called the state space, i.e.,

1) *initial condition:* $\varphi(0, p, x) = x$ for all $p \in P$ and $x \in X$,

2) *cocycle property:*   for all $s, t \in \mathbb{T}^+$, $p \in P$ and $x \in X$,

$$\varphi(s + t, p, x) = \varphi(s, \theta_t(p), \varphi(t, p, x)),$$

3) *continuity:* $(t, p, x) \mapsto \varphi(t, p, x)$ is continuous.

Fig. 3.1   Under the cocyle property the full solution and the concatenated solutions coincide.

Remark 3.2. The base system θ serves as a driving system which makes the cocycle mapping nonautonomous. Skew product flows often have very nice properties when, in particular, the base space P is compact. This occurs when the driving system is, for example, periodic or almost periodic. It provides more detailed information about the dynamical behaviour of the system. George Sell, a pioneering researcher in the area, e.g., see [Sell (1971)], described the effect of a compact base space as being equivalent to *compactifying time*.

3.2.1   *Processes as skew product flows*

Consider a process ϕ on a state space X with time set $\mathbb{T}$, so $\phi : \mathbb{T}_{\geq}^{+} \times X \to X$. For each $t \in \mathbb{T}$ define $\theta_t : \mathbb{T} \to \mathbb{T}$ for each $t \in \mathbb{T}$ by

$$\theta_t(t_0) = t + t_0.$$

The family $\{\theta_t\}_{t \in \mathbb{T}}$ forms a group under addition on $\mathbb{T}$ and is an autonomous dynamical system on $P = \mathbb{T}$.

In addition, define

$$\varphi(t, t_0, x_0) := \phi(t + t_0, t_0, x_0),$$

where $t \in \mathbb{T}^{+}$ is the time that has elapsed since starting at $t_0 \in \mathbb{T}$ in φ, while $t + t_0$ and t_0 are actual times in ϕ.

Remark 3.3. Here and in what follows we use ϕ to denote a process and φ to denote a skew product flow.

The 2-parameter semi-group property

$$\phi(s + t + t_0, t_0, x_0) = \phi\left(s + t + t_0, t + t_0, \phi(t + t_0, t_0, x_0)\right)$$

of the process ϕ becomes the cocycle property

$$\varphi(s + t, t_0, x_0) = \varphi\left(s, \theta_t(t_0), \varphi(t, t_0, x_0)\right)$$

of the skew product flow φ because

$$\varphi(s + t, t_0, x_0) = \phi(s + t + t_0, t_0, x_0)$$

and

$$\varphi(s, \theta_t(t_0), \varphi(t, t_0, x_0)) = \phi\left(s + t + t_0, t + t_0, \phi(t + t_0, t_0, x_0)\right).$$

Here the base space $P = \mathbb{T}$ is locally compact, but not compact.

3.3   Examples

The following examples illustrate how the driving system of the skew product flow formalism often provides more specific, built-in information about the evolution of the system than the process formalism does.

3.3.1   *Autonomous triangular system*

As in Example 3.1 consider the system of scalar ODEs

$$\frac{dp}{dt} = p(1 - p), \qquad \frac{dx}{dt} = -x + p, \tag{3.4}$$

with the initial values

$$p(0) = p_0, \qquad x(0) = x_0,$$

and denote the solutions by
$$p(t) = p(t, p_0), \qquad x(t) = x(t, p_0, x_0).$$
Here t is the time since starting at $t_0 = 0$.

Note that the p-ODE is decoupled from the x-ODE. Its solutions form an autonomous dynamical system on $P = [0, 1]$, but only a semi-dynamical system on $\mathbb{R}^+$ (For $p_0 < 0$ the solutions blow up in finite positive time).

Define $\theta_t(p_0) = p(t, p_0)$ for all $t \in \mathbb{R}$ and $P = [0, 1]$. These form a group under composition on $P = [0, 1]$.

Define $\varphi(t, p_0, x_0) = x(t, p_0, x_0)$. Then φ satisfies the cocycle property on $X = \mathbb{R}$ with respect to θ_t, i.e.,

$$\varphi(s + t, p_0, x_0) = x(s + t, p_0, x_0)$$
$$= x(s, p(t, p_0), x(t, p_0, x_0)) \qquad \text{uniqueness of solutions}$$
$$= \varphi(s, \theta_t(p_0), \varphi(t, p_0, x_0)).$$

The second line is due to the existence and uniqueness of solutions starting at $(\theta_t(p_0), \varphi(t, p_0, x_0))$ at time t.

Note that the base space $P = [0, 1]$ is compact here.

3.3.2 *Driving system need not be an autonomous ODE*

Consider the scalar ODE
$$\frac{dx}{dt} = -x + \cos t.$$
This is like the example in Subsection 3.3.1 with $p(t) = \cos t$. The main difference here is that this $p(t)$ is given directly and not as the solution of a first order ODE.[1]

Nevertheless, we can define a driving system as the shift operator
$$\theta_t(f(\cdot)) := f(t + \cdot), \qquad t \in \mathbb{R},$$
on the function space $(C(\mathbb{R}, \mathbb{R}), \|\cdot\|_\infty)$ of continuous functions $f : \mathbb{R} \to \mathbb{R}$ with the uniform norm
$$\|f\|_\infty := \max_{t \in \mathbb{R}} |f(t)|, \qquad f \in C(\mathbb{R}, \mathbb{R}).$$
Note that the shift operator is an isomorphism on $(C(\mathbb{R}, \mathbb{R}), \|\cdot\|_\infty)$, i.e.,
$$\|\theta_t(f) - \theta_t(g)\|_\infty = \|f(t + \cdot) - g(t + \cdot)\|_\infty$$
$$= \max_{s \in \mathbb{R}} |f(t + s) - g(t + s))|$$
$$= \max_{s \in \mathbb{R}} |f(s) - g(s))| = \|f - g\|_\infty.$$

Lemma 3.1. *The mapping θ_t is continuous in $(C(\mathbb{R}, \mathbb{R}), \|\cdot\|_\infty)$ for each $t \in \mathbb{R}$.*

---

[1]It does, however, satisfy a second order ODE, but that is not relevant here.

Proof. *(Sketch)* This follows directly from the isometry property, i.e.,

$$\|\theta_t(f_n) - \theta_t(f)\|_\infty = \|f_n - f\|_\infty.$$

$\square$

For the base space P we take the closed subset of $C(\mathbb{R}, \mathbb{R})$ consisting of phase shifts of the cosine function, i.e.,

$$P := \bigcup_{0 \le \tau \le 2\pi} \{\cos(\tau + \cdot)\}, \tag{3.5}$$

which is invariant under the shift operator.

This set P is not a linear subspace of $C(\mathbb{R}, \mathbb{R})$, but we can define the induced metric on it by

$$d_P(f, g) := \|f - g\|_\infty, \qquad f, g \in P,$$

so (P, d_P) is a complete metric space. The compactness of (P, d_P) follows from the Ascoli Theorem since P is a closed subset of equi-bounded equi-continuous functions in $(C(\mathbb{R}, \mathbb{R}), \| \cdot \|_\infty)$.

Corollary 3.1. *The mapping θ_t is continuous in (P, d_P) for each $t \in \mathbb{R}$.*

Remark 3.4. In view of the periodicity of the driving system, the parameter set P in (3.5) is homeomorphic to a unit circle. The driving system could alternatively be defined more simply and elegantly on such a circle. The reader should try to show this.

We call the set P defined by (3.5) the *hull* of the function $\cos(\cdot)$. This forms the base space of the driving system $\{\theta_t\}_{t \in \mathbb{R}}$.

Exercise 3.1. Show that $(t, p_0) \mapsto \theta_t(p_0)$ is continuous in $\mathbb{R} \times P$.

The *cocycle mapping* here is given by

$$\varphi(t, p_0, x_0) := x(t, p_0, x_0),$$

the solution mapping of the ODE

$$\frac{dx}{dt} = -x + \theta_t(p_0)$$

with the initial value $x(0) = x_0$, i.e.,

$$\varphi(t, p_0, x_0) = x_0 e^{-t} + e^{-t} \int_0^t e^s \cos(\tau_0 + s)\, ds,$$

where $p_0(\cdot) = \cos(\tau_0 + \cdot)$ corresponding to the phase $\tau_0 \in [0, 2\pi]$, so

$$(\theta_t(p_0))(\cdot) = \theta_t(\cos(\tau_0 + \cdot)) = \cos(\tau_0 + t + \cdot).$$

Exercise 3.2. Show that the mapping $p_0 \mapsto \varphi(t, p_0, x_0)$ continuous. (Hint: The proof depends on uniform continuity and convergence under the integral sign.)

3.3.3 Nonautonomous difference equations

Consider a nonautonomous difference equation on $X = \mathbb{R}$ given by

$$x_{n+1} = f_{j_n}(x_n), \qquad n \in \mathbb{Z}, \tag{3.6}$$

where the functions $f_1, \ldots, f_N$ are given and the j_n are chosen from $\{1, \ldots, N\}$ in some manner.

Specifically, the j_n are the components of a *bi-infinite sequence*

$$\mathbf{s} = (\ldots, j_{-1}, j_0, j_1, j_2, \ldots).$$

Let $P := \{1, \ldots, N\}^{\mathbb{Z}}$ be the collection of all such bi-infinite sequences. Then

$$d_P(\mathbf{s}, \mathbf{s}') := \sum_{n=-\infty}^{\infty} 2^{-|n|} |j_n - j'_n|$$

defines a metric on P.

Lemma 3.2. *(P, d_P) is a compact metric space.*

Define $\theta_n = \theta^n$, the n-fold composition of θ when $n > 0$ and of its inverse θ^{-1} when $n < 0$, where θ is the *left shift operator* of P, i.e.,

$$(\theta(\mathbf{s}))_n = j_{n+1}, \qquad n \in \mathbb{Z}.$$

Lemma 3.3. *$\theta : P \to P$ is Lipschitz continuous in (P, d_P).*

This follows from the definition and

$$d_P(\theta(\mathbf{s}), \theta(\mathbf{s}')) = \sum_{n=-\infty}^{\infty} 2^{-|n|} |j_{n+1} - j'_{n+1}| = \sum_{k=-\infty}^{\infty} 2^{-|k-1|} |j_k - j'_k| \quad \text{with} \quad k = n+1$$

$$\leq 2 \sum_{k=-\infty}^{\infty} 2^{-|k|} |j_k - j'_k| = 2d_P(\mathbf{s}, \mathbf{s}').$$

The *cocycle mapping* for $n \geq 1$ here is

$$\varphi(n, \mathbf{s}, x_0) := f_{j_{n-1}} \circ \cdots \circ f_{j_0}(x_0).$$

It is clear that the mapping $x_0 \mapsto \varphi(n, \mathbf{s}, x_0)$ is continuous due to the continuity of the composition of continuous functions.

What about the continuity of the mapping $\mathbf{s} \mapsto \varphi(n, \mathbf{s}, x_0)$?

Hint: If $\mathbf{s}^{(k)} \to \mathbf{s}$ in (P, d_P), then after a <u>finite</u> k the first n indices $j_l^{(k)} = j_l$ for $l = 0, 1, \ldots, n-1$ for k large enough, i.e., after a finite number of steps the first n indices are all the same.

Remark 3.5. Note that for each index sequence $\mathbf{s} \in P$, the corresponding nonautonomous difference equation (3.6) generates particular process. In contrast, the skew product flow (θ, φ) here includes the dynamics of all such processes. In this sense its dynamics can be considered to be richer.

3.4   Skew product flows as autonomous semi-dynamical systems

Just as a process can be reformulated as an autonomous semi-dynamical system on a product space, so can a skew product flow.

Theorem 3.1. *A skew product flow (θ, φ) on $P \times X$ is an autonomous semi-dynamical system Π on $\mathfrak{X} = P \times X$.*

Proof.   Define $\Pi : \mathbb{T}^+ \times \mathfrak{X} \to \mathfrak{X}$ by

$$\Pi\left(t, (p_0, x_0)\right) := \left(\theta_t(p_0), \varphi(t, p_0, x_0)\right).$$

The *initial condition* property of an semi-dynamical system is obtained by setting $t = 0$ and using the initial condition properties of θ and φ:

$$\Pi\left(0, (p_0, x_0)\right) = \left(\theta_0(p_0), \varphi(0, p_0, x_0)\right) = (p_0, x_0).$$

The *semi-group property* follows from that of θ and the cocycle property of φ:

$$\Pi\left(s + t, (p_0, x_0)\right) = \left(\theta_{s+t}(p_0), \varphi(s + t, p_0, x_0)\right)$$

$$= \left(\theta_s \circ \theta_t(p_0), \varphi(s, \theta_t(p_0), \varphi(t, p_0, x_0))\right)$$

$$= \Pi\left(s, (\theta_t(p_0), \varphi(t, p_0, x_0))\right) = \Pi\left(s, \Pi\left(t, (p_0, x_0)\right)\right).$$

Finally, the *continuity* of the mapping

$$(t, p_0, x_0) \mapsto \Pi\left(t, (p_0, x_0)\right) = \left(\theta_t(p_0), \varphi(t, p_0, x_0)\right)$$

is inherited from that of its constituent functions. $\square$

Remark 3.6. Unlike the semi-dynamical system on the product space $\mathfrak{X} = \mathbb{T} \times X$ generated by a process, the semi-dynamical system on the product space $\mathfrak{X} = P \times X$ generated by a skew product flow can have compact invariant sets, in particular when the base space P is compact. In this case the product semi-dynamical system may be able to provide useful information.

Chapter 4

Entire solutions and invariant sets

Invariant sets and bounded entire solutions provide much useful information about the dynamical behaviour of a dynamical system. Entire solutions are sometimes called *complete* or *full* solutions in the literature.

4.1 Autonomous case

Consider an autonomous semi-dynamical system π on a complete metric state space (X, d_X) with the time set $\mathbb{T}$.

Definition 4.1. An *entire solution* of a semi-dynamical system π on a state space X with the time set $\mathbb{T}$ is a mapping $e : \mathbb{T} \to X$ with the property that

$$e(t) = \pi(t - s, e(s)) \qquad (4.1)$$

for all s, $t \in \mathbb{T}$ with $t \geq s$.

Note that $t - s \in \mathbb{T}^+$, the time set on which the semi-dynamical system π is defined. However, the entire solution e is defined for all $t \in \mathbb{T}$, not just in $\mathbb{T}^+$.

Example 4.1. Consider the continuous time semi-dynamical system on $X = \mathbb{R}$ defined by the solutions of the initial value problem

$$\frac{dx}{dt} = -x^3, \quad x(0) = x_0, \qquad (4.2)$$

i.e.,

$$x(t, x_0) = \frac{x_0}{\sqrt{1 + 2x_0^2 t}} \quad \text{for all } t > -\frac{1}{2x_0^2}.$$

These solutions exist only on the semi-infinite interval $(-\frac{1}{2x_0^2}, \infty)$. The time $t = -\frac{1}{2x_0^2}$ is a vertical asymptote.

The only entire solution is $e(t) \equiv 0$.

The semi-dynamical system here is defined by $\pi(t, x_0) = x(t, x_0)$ for all $t \geq 0$ and $x_0 \in \mathbb{R}$ (even though all solutions can be extended backwards at least for a short time depending on the initial value x_0.

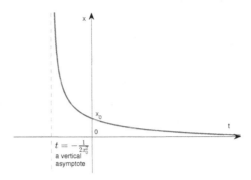

Fig. 4.1   Typical solution of the ODE (4.2).

It is clear that steady state, i.e., constant solutions are entire solutions, but they are not the only possibility.

Example 4.2. The initial value problem

$$\frac{dx}{dt} = 4x(1-x), \quad x(0) = x_0, \tag{4.3}$$

has the explicit solution

$$x(t, x_0) = \frac{x_0 e^{4t}}{1 - x_0 + x_0 e^{4t}}.$$

Its steady state solutions are $\bar{x}(t) \equiv 0$ and $\bar{x}(t) \equiv 1$ for all $t \in \mathbb{R}$. These are obviously entire solutions. In addition all solutions with $x_0 \in (0,1)$ exist for all time, so are also entire solutions.

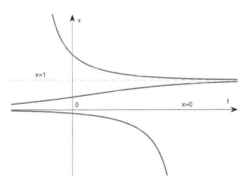

Fig. 4.2   Typical solutions of the ODE (4.3).

For $x_0 > 1$ the solutions exist for all $t \geq 0$, but have a vertical asymptote for a *finite* negative time depending on x_0. Similarly, for $x_0 < 0$ the solutions exist for all $t \leq 0$ but have a vertical asymptote at a finite positive time that depends on x_0.

Note that the solution mapping $x(t, x_0)$ in the above ODE defines an autonomous semi-dynamical system (semi-group) on the state space $X = \mathbb{R}^+$, but not on the whole of $\mathbb{R}$. However, it defines an autonomous dynamical system (group) on $X = [0, 1]$.

The steady state solutions of a semi-dynamical system, i.e., with $\pi(t, \bar{x}) = \bar{x}$ for all $t \in \mathbb{R}$, are obvious examples of entire solutions. The singleton subset $A = \{\bar{x}\}$ then satisfies $\pi(t, \{\bar{x}\}) = \{\bar{x}\}$ for all $t \in \mathbb{T}^+$. This is a special case of the following definition.

Definition 4.2. A subset A of X is called *invariant* (with respect to a semi-dynamical system π) if

$$\pi(t, A) = A \quad \text{for all } t \in \mathbb{T}^+,$$

where

$$\pi(t, A) := \bigcup_{a \in A} \{\pi(t, a)\}.$$

Thus π maps A *onto* itself for each $t \in \mathbb{T}^+$. The set $A = [0, 1]$ in Example 4.2 is example of such a set.

There is a weaker version of this property, where π just maps A *into* itself.

Definition 4.3. A subset A of X is called *positively invariant* (with respect to a semi-dynamical system π) if

$$\pi(t, A) \subset A \qquad \text{for all } t \in \mathbb{T}^+.$$

Example 4.3. The set $A = [0, 1]$ in Example 4.2 is invariant, while the sets $A = [0, 1 + r]$ for $r > 0$ are only positively invariant.

Invariant sets are made up of entire solutions.

Lemma 4.1. *Let A be an invariant set w.r.t. a semi-dynamical system π. Then for every $a \in A$ there exists an entire solution $e_a : \mathbb{T} \to A$ with $e_a(0) = a$.*

Proof: Invariant means that $\pi(t, A) = A$ for all $t \in \mathbb{T}^+$. In particular, given $a \in A$, then $\pi(t, a) \in A$ for $t \in \mathbb{T}^+$. Define $e_a(t) := \pi(t, a)$ for $t \in \mathbb{T}^+$.

In particular, $\pi(1, A) = A$, so there exists an $a_{-1} \in A$ such that

$$\pi(1, a_{-1}) = a.$$

Define $e_a(t) := \pi(1 + t, a_{-1})$ for $t \in [-1, 0] \cap \mathbb{T}$. Obviously, then

$$e_a(-1) = \pi(0, a_{-1}) = a_{-1}, \qquad e_a(0) = \pi(1, a_{-1}) = a$$

and

$$e_a(t) = \pi(1 + t, a_{-1}) \in \pi(1 + t, A) = A$$

for all $t \in [-1, 0] \cap \mathbb{T}$.

We repeat this argument for $t \in [-n-1, -n] \cap \mathbb{T}$ and construct a mapping $e_a :$ $\mathbb{T} \to A$ with $e_a(0) = a$ and $e_a(t) \in A$ for all $t \in \mathbb{T}$.

It remains to show that $e_a(t)$ satisfies the relationship (4.1). This follows from the definition of $e_a(t)$ and the semi-group property of π. $\qquad\square$

Invariant subsets are important as most of the long term dynamics is characterised by them.

4.2　Nonautonomous case

Now consider a 2-parameter semi-group or process ϕ on the metric state space X and time set $\mathbb{T}$.

An entire solution of a process is defined analogously to an entire solution of an autonomous semi-dynamical system.

Definition 4.4. An *entire solution* of a process ϕ on a state space X with the time set $\mathbb{T}$ is a mapping $e : \mathbb{T} \to X$ with the property that

$$e(t) = \phi(t, t_0, e(t_0))$$

for all $(t, t_0) \in \mathbb{T}^2_{\geq} := \{(t, t_0) \in \mathbb{T} \times \mathbb{T} : t \geq t_0\}$.

Steady state solutions are entire solutions, but there may be other interesting *bounded* entire solutions.

Example 4.4. All solutions of the ODE

$$\frac{dx}{dt} = -x + \cos t$$

are entire solutions, but the only *bounded* entire solution is

$$\bar{x}(t) = \frac{1}{2} \cos t + \frac{1}{2} \sin t.$$

Moreover, $\bar{x}(t)$ is a *periodic* solution, in fact, the only periodic solution. This ODE has no steady state solutions.

Nonautonomous invariant sets

The analogue of an invariant set for an autonomous semi-dynamical system is *too restrictive* for a process, i.e., a subset A of X such that

$$\phi(t, t_0, A) = A \qquad \text{for all } (t, t_0) \in \mathbb{T}^2_{\geq}.$$

Such a subset contains steady state solutions if there are any, but excludes almost everything else such as the periodic solution in the previous example.

How then should we define an invariant set for a nonautonomous process?

For this it is useful to consider the autonomous semi-dynamical system Π on $\mathfrak{X} = \mathbb{T} \times X$ defined in terms of a process ϕ on X with time set $\mathbb{T}$ (see Section 2.4 of Chapter 2) by

$$\Pi(\tau, (t_0, x_0)) = (\tau + t_0, \phi(\tau + t_0, t_0, x_0)).$$

Proposition 4.1. *Let $\mathfrak{A} \subset \mathfrak{X} = \mathbb{T} \times X$ be a nonempty invariant set of the autonomous semi-dynamical system Π on $\mathfrak{X}$, i.e., $\Pi(\tau, \mathfrak{A}) = \mathfrak{A}$ for all $\tau \in \mathbb{T}^+$.*
Then $\mathfrak{A} = \bigcup_{t \in \mathbb{T}} \{t\} \times A_t$, where A_t is a nonempty subset of X for each $t \in \mathbb{T}$.

Proposition 4.2. *Let A_t, $t \in \mathbb{T}$, be the nonempty subsets of X given in Proposition 4.1. Then*

$$\phi(t, t_0, A_{t_0}) = A_t \quad \text{for all } (t, t_0) \in \mathbb{T}_{\geq}^2.$$

We will prove these Propositions at the end of the section. Proposition 4.2 gives us a hint how to define an invariant set in the nonautonomous case.

Definition 4.5. A family $\mathcal{A} = \{A_t, t \in \mathbb{T}\}$ of nonempty subsets A_t of X is called *invariant* with respect to a process or ϕ-invariant if

$$\phi(t, t_0, A_{t_0}) = A_t \quad \text{for all } (t, t_0) \in \mathbb{T}_{\geq}^2.$$

For example, for a steady state solution, i.e., with $\phi(t, t_0, \bar{x}) \equiv \bar{x}$ for all $(t, t_0) \in \mathbb{T}_{\geq}^2$, the family of identical singleton subsets $A_t \equiv \{\bar{x}\}$ for all $t \in \mathbb{T}$ is ϕ-invariant. In Example 4.4, the family of singleton sets

$$A_t = \left\{ \frac{1}{2} \cos t + \frac{1}{2} \sin t \right\}, \qquad t \in \mathbb{R},$$

formed by the periodic solution is ϕ-invariant.

The proof of the following Lemma is analogous to that of Lemma 4.1 in the autonomous case.

Lemma 4.2. *Let $\mathcal{A} = \{A_t, t \in \mathbb{T}\}$ be a ϕ-invariant family of subsets of X. Then for any $a_0 \in A_{t_0}$ and any $t_0 \in \mathbb{T}$ there exists an entire solution $e_{a_0,t_0} : \mathbb{T} \to X$ of ϕ such that $e_{a_0,t_0}(t_0) = a_0$ and*

$$e_{a_0,t_0}(t) \in A_t \quad \text{for all } t \in \mathbb{T}.$$

Proof of Proposition 4.1

The set $\mathfrak{A} \subset \mathfrak{X} = \mathbb{T} \times X$ is nonempty by assumption, so there is at least one point $(t_0, a_0) \in \mathfrak{A}$, which means that $a_0 \in A_{t_0}$. Hence A_{t_0} is nonempty.

Then A_{t_1} is nonempty for every $t_1 > t_0$, since $\phi(t_1, t_0, a_0) \in A_{t_1}$. This follows by Π-invariance of $\mathfrak{A}$, i.e., $\Pi(t_1 - t_0, \mathfrak{A}) = \mathfrak{A}$ and the fact that

$$\Pi(t_1 - t_0, (t_0, a_0)) = (t_0 + (t_1 - t_0), \phi(t_0 + (t_1 - t_0), t_0, a_0))$$

$$= (t_1, \phi(t_1, t_0, a_0)) \in \mathfrak{A}$$

so $\phi(t_1, t_0, a_0) \in A_{t_1}$. Hence A_{t_1} is nonempty for $t_1 > t_0$.

Now for $t_1 < t_0$, by invariance $\Pi(t_0 - t_1, \mathfrak{A}) = \mathfrak{A}$, so there exists a $(t^*, a^*) \in \mathfrak{A}$ such that

$$\Pi(t_0 - t_1, (t^*, a^*)) = (t_0, a_0),$$

i.e., $t_0 - t_1 + t^* = t_0$ and $\phi(t_0 - t_1 + t^*, t^*, a^*) = a_0$, which means that $t^* = t_1$ and $\phi(t_0, t^*, a^*) = a_0$. Hence $a^* \in A_{t_1}$, so A_{t_1} is nonempty for $t_1 < t_0$. $\square$

Proof of Proposition 4.2

Let $t_1 > t_0$ and $a_0 \in A_{t_0}$. By the invariance $\Pi(t_1 - t_0, \mathfrak{A}) = \mathfrak{A}$ we have

$$\Pi(t_1 - t_0, (t_0, a_0)) = (t_1, \phi(t_1, t_0, a_0)) \in (t_1, A_{t_1}),$$

which means that $\phi(t_1, t_0, a_0) \in A_{t_1}$. Hence $\phi(t_1, t_0, A_{t_0}) \subseteq A_{t_1}$.

Now let $a_1 \in A_{t_1}$. By the onto nature of the above invariance there exists a point $(t^*, a^*) \in \mathfrak{A}$ such that $\Pi(t_1 - t_0, (t^*, a^*)) = (t_1, a_1)$ or

$$(t_1 - t_0 + t^*, \phi(t_1 - t_0 + t^*, t^*, a^*)) = (t_1, a_1),$$

which means that $t_1 - t_0 + t^* = t_1$ and hence that $t_0 = t^*$. In addition,

$$\phi(t_1 - t_0 + t^*, t^*, a^*) = \phi(t_1, t_0, a^*) = a_1.$$

Obviously, $a^* \in A_{t^*} = A_{t_0}$ since $t^* = t_0$. This means that $\phi(t_1, t_0, A_{t_0}) \supseteq A_{t_1}$.

Combining both directions gives

$$\phi(t, t_0, A_{t_0}) = A_t \qquad \text{for all } (t, t_0) \in \mathbb{T}^2_{\geq}.$$

$\square$

PART 2
Pullback attractors

Chapter 5

Attractors

Let (X, d_X) be a complete metric space and let $\mathcal{H}(X)$ be the collection of all *nonempty compact subsets* of X.

The distance between two points $a,\ b \in X$ is given by
$$d_X(a, b) = d_X(b, a) \qquad \text{(symmetric!)}.$$
We define the distance between a point $a \in X$ and a nonempty compact subset B in X by
$$\operatorname{dist}_X(a, B) := \inf_{b \in B} d_X(a, b).$$

Remark 5.1. The mapping $b \mapsto d(a, b)$ is continuous for a fixed, in fact
$$|d_X(a, b) - d_X(a, b')| \leq d(b, b'),$$
and the subset B is nonempty and compact, so the inf can be replaced by min here, i.e., it is actually attained.

Then we define the distance of a compact subset A from a compact subset B by
$$\operatorname{dist}_X(A, B) := \sup_{a \in A} \operatorname{dist}_X(a, B) = \sup_{a \in A} \inf_{b \in B} d_X(a, b),$$
which is sometimes written $H_X^*(A, B)$ and called the *Hausdorff separation* or semi-distance of A f rom B.

Remark 5.2. The function $a \mapsto \operatorname{dist}_X(a, B)$ is continuous for fixed B and the set A is compact, so the sup here can be replaced by max.

The Hausdorff separation, $\operatorname{dist}_X(A, B)$ satisfies the *triangle inequality*
$$\operatorname{dist}_X(A, B) \leq \operatorname{dist}_X(A, C) + \operatorname{dist}_X(C, B).$$
However, $\operatorname{dist}_X(A, B)$ is *not* a metric, since it can be equal to zero without the sets being equal, i.e., $\operatorname{dist}_X(A, B) = 0$ if $A \subset B$.

Define
$$H_X(A, B) := \max\left\{\operatorname{dist}_X(A, B), \operatorname{dist}_X(B, A)\right\}.$$
This is a metric on $\mathcal{H}(X)$ called the *Hausdorff metric*.

Theorem 5.1. $(\mathcal{H}(X), H_X)$ *is a complete metric space.*

5.1 Attractors of autonomous systems

Let π be an autonomous semi-dynamical system on a complete metric space (X, d_X) with time set $\mathbb{T}^+$.

To facilitate the comparison with attractors of nonautonomous systems, which we introduce below, we recall some definitions and results about autonomous attractors from Chapter 1.

Definition 5.1. A nonempty compact subset A of X is called an *attractor* of the semi-dynamical system π if

- i) π-invariant A is π-invariant, i.e., $\pi(t, A) = A$ for all $t \in \mathbb{T}^+$,

- ii) A attracts nonempty bounded subsets of X by π, i.e.,

$$\text{dist}_X(\pi(t, D), A) \to 0 \quad \text{as } t \to \infty$$

for all bounded subsets D of X.

Example 5.1. Consider the semi-dynamical system generated by the ODE on $\mathbb{R}^1$

$$\frac{dx}{dt} = x\left(1 - x^2\right), \tag{5.1}$$

which has steady state solutions 0 and ± 1. Its attractor $A = [-1, +1]$.

Attractors may have a very complicated geometry and be difficult to determine clearly, e.g., a fractal set. More convenient to describe and determine is an *absorbing set*, which usually has simple geometry such as a box or sphere.

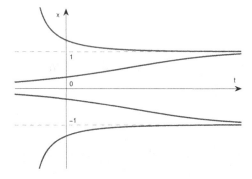

Fig. 5.1   Typical solutions of the ODE (5.1).

Definition 5.2. A nonempty subset B of X is called an absorbing set for an semi-dynamical system π if for every nonempty bounded subset D of X there exists a finite time $T_D \geq 0$ such that

$$\pi(t, D) \subseteq B \quad \text{for all } t \geq T_D.$$

Example 5.2. In Example 5.1 the attractor is $A = [-1, +1]$ and any subset $B_r = [-1 - r, 1 + r]$ with $r > 0$ is an absorbing set.

Example 5.3. The Lorenz equations in $\mathbb{R}^3$ are famous in meteorology as a system of ODEs with a strange chaotic attractor. They are given by

$$\frac{dx}{dt} = -\sigma x + \sigma y, \qquad \frac{dy}{dt} = -xz + rx - y, \qquad \frac{dz}{dt} = xy - bz \qquad (5.2)$$

with parameters σ, r, $b > 0$.

Consider the Lyapunov-like function

$$V(x, y, z) = \frac{1}{2} \left[x^2 + y^2 + (z - \sigma - r)^2 \right].$$

Then along solutions of the system of ODEs (5.2)

$$\frac{d}{dt} V(x(t), y(t), z(t)) = x(t) \left(-\sigma x(t) + \sigma y(t) \right) + y(t) \left(-x(t)z(t) + rx(t) - y(t) \right)$$

$$+ red(z(t) - \sigma - r) \left(x(t)y(t) - bz(t) \right)$$

$$= -\sigma x(t)^2 - y(t)^2 - \frac{1}{2} b \left(z(t) - \sigma - r \right)^2 - red\frac{1}{2} bz(t)^2 + \frac{1}{2} b(\sigma + r)^2.$$

Let $\lambda := \min \left\{ \sigma, 1, \frac{1}{2} b \right\}$. Then

$$\frac{d}{dt} V(x(t), y(t), z(t)) \leq -2\lambda V(x(t), y(t), z(t)) + \frac{1}{2} b(\sigma + r)^2.$$

We can integrate this differential inequality to obtain

$$V(x(t), y(t), z(t)) \leq V(x_0, y_0, z_0)e^{-2\lambda t} + \frac{b}{4\lambda}(\sigma + r)^2 \left(1 - e^{-2\lambda t} \right)$$

$$\leq 1 + \frac{b}{4\lambda}(\sigma + r)^2 =: R^*$$

for $t \geq T_D := \frac{1}{2\lambda} \ln V_D$, where $V_D := \max_{(x_0, y_0, z_0) \in D} V(x_0, y_0, z_0)$.

This means that after time T_D the value of $V(x(t), y(t), z(t))$ will be bounded by R^*. This forms a closed and bounded ball in $\mathbb{R}^3$, i.e., an absorbing set for the Lorenz equations.

If we know that there is an absorbing compact set, then there is an attractor inside this absorbing set. We recall Theorem 1.3.

Theorem 5.2. *Suppose that the compact set B is an absorbing set for a semi-dynamical system π. Then π has an attractor in B given by*

$$A = \bigcap_{t \geq 0} \overline{\bigcup_{s \geq t} \pi(s, B)}.$$

If, in addition, the absorbing set B is positively invariant, i.e., $\pi(t, B) \subset B$ for all $t \geq 0$, then

$$A = \bigcap_{t \geq 0} \pi(t, B).$$

If the absorbing compact set B is not positively invariant and the state space $X = \mathbb{R}^d$, then we can find a compact larger set B^* which is both absorbing and positively invariant, namely

$$B^* := \bigcup_{0 \leq t \leq T_B} \pi(t, B) = \pi([0, T_B], B),$$

where T_B is the time for π to absorb the set B into itself, i.e.,

$$\pi(T_B, B) \subseteq B.$$

Firstly, note that B^* is compact as the continuous image of the compact subset $[0, T_B] \times B$ in $\mathbb{R} \times \mathbb{R}^d$.

Secondly, B^* is an absorbing set because of its definition and the absorbing property of B, i.e., for any bounded set D of $\mathbb{R}^d$ there is a T_D such that

$$\pi(T_D, D) \subseteq B \subseteq B^*.$$

Proposition 5.1. B^* *is π-positively invariant.*

Proof. Let $b^* \in B^*$. Then there exists a $t \in [0, T_D]$ and a $b_t \in B$ such that $b^* = \pi(t, b_t)$. Then for any $s \in \mathbb{T}^+$,

$$\pi(s, b^*) = \pi(s, \pi(t, b_t)) = \pi(s + t, b_t),$$

where $\pi(s + t, b_t) \in B^*$ if $s + t \leq T_B$ by the definition of B^* and $\pi(s + t, b_t) \in B$ if $s + t \geq T_B$ by the absorbing property of B. Hence,

$$\pi(s, b^*) \in B^* \quad \text{for all} \quad s \in \mathbb{T}^+.$$

However, $b^* \in B^*$ was chosen arbitrarily, so $\pi(s, B^*) \subseteq B^*$ for all $s \in \mathbb{T}^+$.   □

Remark 5.3. The above construction also holds in a general complete metric space X, when the positive invariant set B is just closed and bounded. Then the invariant set B^* is also closed and bounded. As mentioned in Remark 1.5 some additional compactness property of π is then needed to obtain the existence of an attractor.

5.2   What is a nonautonomous attractor?

Let ϕ be a process on X with the time set $\mathbb{T}$. It is fairly clear that an attractor for ϕ should be a family $\mathcal{A} = \{A_t, t \in \mathbb{T}\}$ of nonempty compact subsets A_t of X, which is ϕ-invariant, i.e.,

$$\phi(t, t_0, A_{t_0}) = A_t \quad \text{for all } (t, t_0) \in \mathbb{T}_2^+.$$

There is, however, a problem with convergence.

Example 5.4. The scalar ODE

$$\frac{dx}{dt} + x = \cos t \tag{5.3}$$

has no steady state solutions.

We can use an integrating factor $\mu(t) := e^t$ to obtain

$$\frac{d}{dt}\left(e^t x(t)\right) = e^t \cos t. \tag{5.4}$$

Then we use complex integration and take the real part $\Re$:

$$\int e^t \cos t\, dt = \Re \int e^{(1+i)t}\, dt = \frac{1}{2}e^t\left(\cos t + \sin t\right).$$

Integrating equation (5.4) from t_0 to t gives

$$x(t)e^t - x(t_0)e^{t_0} = \frac{1}{2}e^t\left(\cos t + \sin t\right) - \frac{1}{2}e^{t_0}\left(\cos t_0 + \sin t_0\right),$$

so the solution with initial value $x(t_0) = x_0$ is

$$x(t, t_0, x_0) = \left[x_0 - \frac{1}{2}\left(\cos t_0 + \sin t_0\right)\right]e^{-(t-t_0)} + \frac{1}{2}\left(\cos t + \sin t\right). \tag{5.5}$$

There is *no* limit as $t \to \infty$ because $\frac{1}{2}\left(\cos t + \sin t\right)$ is oscillating in t.

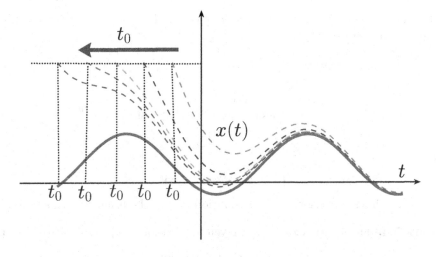

Fig. 5.2   Pullback attraction.

Instead, let us take the *pullback limit*, i.e., $t_0 \to -\infty$ with t held fixed. Then the term $\frac{1}{2}\left(\cos t + \sin t\right)$ is like a constant (i.e., w.r.t. t_0) and the other terms are dominated by

$$e^{-(t-t_0)} = e^{-t+t_0} \to 0 \quad \text{as } t_0 \to -\infty \quad \text{(for fixed } t\text{)}.$$

(In fact, also for $t \to \infty$ with t_0 held fixed.)

In this way we obtain the *pullback limit*

$$\bar{x}(t) = \frac{1}{2}\left(\cos t + \sin t\right).$$

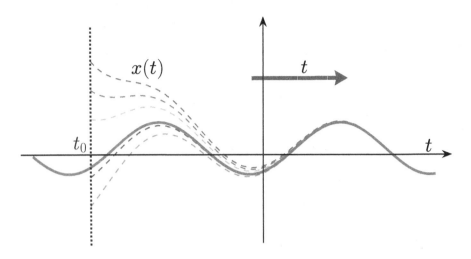

Fig. 5.3   Forward attraction.

It is easy to show that $\bar{x}(t)$ is a solution of the ODE. Moreover, from the explicit solution above, we have

$$|x(t) - \bar{x}(t)| = \left| x_0 - \frac{1}{2}\left(\cos t_0 + \sin t_0\right) \right| e^{-(t-t_0)}$$

$$\leq \left(|x_0| + \frac{1}{2}\left(|\cos t_0| + |\sin t_0|\right) \right) e^{-(t-t_0)}$$

$$\leq \left(|x_0| + 1 \right) e^{-(t-t_0)}$$

$$\to 0 \qquad \text{as} \quad t \to \infty \quad \text{with fixed } t_0.$$

Thus we also have convergence to the solution $\bar{x}(t)$ in the usual forwards sense.

Remark 5.4. The solution $x(t, t_0, x_0)$ given in equation (5.5) satisfies the property

$$x(t + \tau, t_0 + \tau, x_0) = x(t, t_0, x_0) \quad \text{for all } t \geq t_0, x_0,$$

where $\tau = 2\pi$ is the period for the driving process $\cos t$. It is an example of what Haraux [Haraux (1991)] called a *periodic process*. It does not mean that the solutions are τ-periodic, but rather that they are invariant under a τ-time shift for this specific τ. This is in contrast with an autonomous system, for which the solutions are invariant under all time translations.

The previous example introduces two types of convergence. It is better to write the solution as $x(t) = x(t, t_0, x_0)$ here.

Definition 5.3. Forward convergence (moving target) positively

$$\lim_{t \to \infty} |x(t, t_0, x_0) - \bar{x}(t)| = 0 \qquad t_0 \text{ held fixed.}$$

Definition 5.4. Pullback convergence (fixed target)

$$\lim_{t_0 \to -\infty} |x(t, t_0, x_0) - \bar{x}(t)| = 0 \qquad t \text{ held fixed.}$$

Remark 5.5. Pullback convergence uses information about the system from the past, whereas forward convergence uses information about the system in the future.

What is the relationship between pullback and forward convergence?

Are they independent concepts?

In an autonomous system they are equivalent, since $t - t_0 \to \infty$ if either $t \to \infty$ with t_0 held fixed or $t_0 \to -\infty$ with t held fixed.

Chapter 6

Nonautonomous equilibrium solutions

The concept of pullback convergence introduced in the previous chapter provides a new tool which allows special kinds of entire solutions of nonautonomous ordinary differential equations (ODEs) to be constructed and their asymptotic behaviour to be characterised. Instead of the usual steady state equilibria of autonomous ordinary differential equations there are now time varying nonautonomous equilibria that are entire solutions determined by pullback attraction. These are special cases of more general nonautonomous pullback attractors and provide insight into how such attractors should be defined.

They are introduced and illustrated here in terms the nonautonomous SIR equations in epidemiology. Autonomous SIR models have been investigated extensively in the literature, where steady state solutions play an important role in the analysis. However, given the seasonal variations in many diseases, the nonautonomous version of these equations are often more realistic. This requires the inclusion of time variable coefficients or forcing terms in the models. As a consequence steady state solutions may no longer exists. Nonautonomous equilibria are the appropriate counterpart. These will be investigate in detail in this chapter in the context of the simpler SI models since the nonautonomous equilibrium solutions can be determined explicitly in analytical form.

6.1 The SIR model

The SIR equations have variables S (susceptibles), I (infected) and R (recovered). They will be investigated here with time variable parameters or forcing (deterministic or random) in the model, so the total population $N(t) = S(t) + I(t) + R(t)$ need not be constant.

This will be achieved first through a temporal forcing term given by a continuous function $q : \mathbb{R} \to \mathbb{R}$ taking positive bounded values, i.e.,

$$q(t) \in [q^-, q^+] \quad \text{for all t } \in \mathbb{R},$$

where $0 < q^- \leq q^+$, and later with a time variable interaction coefficient γ. In general, the time variability could be arbitrary but in typical situations they are periodic or almost periodic functions.

The SIR model is given by the system of ODEs

$$\frac{dS}{dt} = aq(t) - aS + bI - \gamma\frac{SI}{N}, \tag{6.1}$$

$$\frac{dI}{dt} = -(a+b+c)I + \gamma\frac{SI}{N}, \tag{6.2}$$

$$\frac{dR}{dt} = cI - aR, \tag{6.3}$$

where the parameters a, b, c and γ are positive constants. This means that solutions with nonnegative initial values remains nonnegative. (Later the parameters will also be allowed to vary in time within suitable positive bounds).

The system (6.1) has no steady state solutions when $q(t)$ is not identically equal to a constant.

Adding both sides of the ODEs gives the scalar nonautonomous ODE

$$\frac{dN}{dt} = a(q(t) - N), \tag{6.4}$$

which has the solution

$$N(t) = N_0 e^{-a(t-t_0)} + ae^{-at}\int_{t_0}^{t} q(s)e^{as}\, ds.$$

Now

$$ae^{-at}\int_{t_0}^{t} e^{as}\, ds = 1 - e^{-a(t-t_0)},$$

so given the positivity bounds on $q(t)$ one sees that the integral takes values between

$$q^{-}\left(1 - e^{-a(t-t_0)}\right) \leq ae^{-at}\int_{t_0}^{t} q(s)e^{as}\, ds \leq q^{+}\left(1 - e^{-a(t-t_0)}\right).$$

Hence the total population is bounded above and below, specifically

$$q_{-} + (N_0 - q_{-})\, e^{-a(t-t_0)} \leq N(t) \leq q^{+} + \left(N_0 - q^{+}\right) e^{-a(t-t_0)}.$$

It follows from these inequalities that the simplex slab

$$\Sigma_3^{\pm} = \left\{(S,I,R) \ : \ S,I,R \geq 0, \ N = S+I+R \in [q^{-}, q^{+}]\right\} \subset \mathbb{R}^3$$

attracts all populations starting outside it and that populations starting within it remain there. When the forcing term $q(t)$ is not identically equal to a constant, the simplex slab $\Sigma_3^{\pm}$ will in fact absorb outside populations in a finite time and will be positive invariant (rather than strictly invariant). Thus one thus restrict attention to the dynamics within $\Sigma_3^{\pm}$.

The ODE (6.4) has no steady state solutions, but it has what Chueshov [Chueshov (2002)] called a *nonautonomous equilibrium solution*, which is determined by taking the *pullback limit*, i.e., as $t_0 \to -\infty$ with t held fixed, namely,

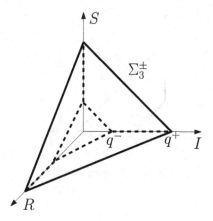

Fig. 6.1   Simplex slap $\Sigma_3^{\pm}$.

$$\widehat{N}(t) = ae^{-at} \int_{-\infty}^{t} q(s)e^{as}\,ds. \qquad (6.5)$$

In this example the nonautonomous equilibrium solution also forward attracts all other solution of the ODE (6.4), i.e.

$$\left| N(t) - \widehat{N}(t) \right| = e^{-a(t-t_0)} \left| N(t_0) - \widehat{N}(t_0) \right| \to 0 \quad \text{as} \quad t \to \infty.$$

The nonautonomous equilibrium solution $\widehat{N}(t)$ of the scalar nonautonomous ODE (6.4), which was constructed by pullback convergence, is thus asymptotically Lyapunov stable.

6.2   The SI equations with variable population

For simplicity assume now that the R term is not present and $c = 0$, since then the solutions can be determined explicitly. The SIR equations (6.1) reduce to

$$\frac{dS}{dt} = aq(t) - aS + bI - \gamma\frac{SI}{N}, \qquad (6.6)$$

$$\frac{dI}{dt} = -(a+b)I + \gamma\frac{SI}{N}. \qquad (6.7)$$

Now $I = 0$ is a steady state of (6.7), which means that the S face of the polygonal set

$$\Sigma_2^{\pm} = \{(S,I) \,:\, S, I \geq 0,\, N = S + I \in [q^-, q^+]\} \subset \mathbb{R}^2$$

is invariant and the equations on it reduce to

$$\frac{dS}{dt} = aq(t) - aS. \qquad (6.8)$$

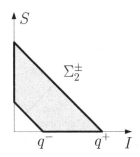

Fig. 6.2   Simplex slap $\Sigma_3^{\pm}$.

This has the explicit solution

$$S(t) = S_0 e^{-a(t-t_0)} + ae^{-at} \int_{t_0}^t q(s)e^{as}\, ds.$$

There no steady state solution when $q(t)$ is not a constant, but there is a nonautonomous equilibrium solution found be taking the pullback limit (i.e., as $t_0 \to -\infty$ with t held fixed), namely,

$$\widehat{S}_1(t) = ae^{-at} \int_{-\infty}^t q(s)e^{as}\, ds.$$

This also forward attracts all other solution of the S equation on the $I = 0$ face, i.e.

$$\left| S(t) - \widehat{S}_1(t) \right| \to 0 \quad \text{as} \quad t \to \infty.$$

In particular, the SI dynamics has no steady state but a nonautonomous equilibrium solution $(\widehat{S}_1(t), 0)$.

Lemma 6.1. *The nonautonomous equilibrium solution $(\widehat{S}_1(t), 0)$ is globally asymptotically stable in $\Sigma_2^{\pm}$ when $\gamma \leq a + b$. It is unstable when $\gamma > a + b$.*

Proof. Suppose that $\gamma < a + b$. Multiply the I equation by $2I$ to obtain

$$\frac{d}{dt}I^2 = -2(a+b)I^2 + 2\gamma\frac{SI^2}{N} \leq 2(\gamma - a - b)I^2$$

because $I^2 \geq 0$ and

$$0 \leq \frac{S}{N} \leq 1.$$

Hence

$$I^2(t) \leq I^2(0)e^{-2|\gamma-a-b|t} \to 0 \quad \text{as} \quad t \to \infty.$$

If $\gamma = a + b$, the I equation gives

$$\frac{d}{dt}I = -(a+b)I + \gamma\frac{SI}{N} = -(a+b)I + \gamma\frac{(N-I)I}{N}$$

$$= (\gamma - a - b)I - \gamma\frac{I^2}{N}$$

$$= -\gamma\frac{I^2}{N} < -\gamma\frac{I^2}{q^-} < 0$$

for all $I > 0$, so

$$0 \leq I(t) \leq \frac{I(0)q^-}{q^- + \gamma I(0)t} \to 0 \quad \text{as} \quad t \to \infty.$$

In both cases

$$S(t) - \widehat{S}_1(t) = \left(N(t) - \widehat{N}(t)\right) - I(t) \to 0 \quad \text{as} \quad t \to \infty,$$

since $\widehat{S}_1(t) \equiv \widehat{N}(t)$ and both terms on the right converge to zero with increasing time. The nonautonomous equilibrium solution is thus globally asymptotically stable when $\gamma \leq a + b$.

Finally, let $\gamma > a+b$ and suppose that $I \leq \varepsilon N$ in the I equation, so $S \geq (1-\varepsilon)N$. Then

$$\frac{d}{dt}I = -(a+b)I + \gamma\frac{SI}{N} \geq -(a+b)I + \gamma(1-\varepsilon)I \geq (\gamma(1-\varepsilon) - a - b)I,$$

so $I(t)$ is strictly increasing if $0 < I \leq \varepsilon N$, provided $\varepsilon < 1 - \frac{a+b}{\gamma}$. In particular, the nonautonomous equilibrium solution $(\widehat{S}_1(t), 0)$ is unstable when $\gamma > a + b$. $\qquad\square$

6.2.1 *Nontrivial limiting solutions*

Now consider the possibility of a nontrivial equilibrium, i.e., $\widehat{I} \neq 0$. The limiting population will be used here, $\widehat{N} = \widehat{S} + \widehat{I}$, in the SI equations (6.6)–(6.7). Then the equation for I can be rewritten

$$\frac{d\widehat{I}}{dt} = -(a+b)\widehat{I} + \gamma\frac{I}{\widehat{N}}(\widehat{N} - \widehat{I}),$$

that is as

$$\frac{d\widehat{I}}{dt} = (\gamma - a - b)\widehat{I} - \gamma\frac{1}{\widehat{N}}\widehat{I}^2.$$

This is a Bernoulli equation which can be solved with the substitution $V = \widehat{I}^{-1}$, to give the linear ODE

$$\frac{dV}{dt} + (\gamma - a - b)V = \gamma\frac{1}{\widehat{N}}.$$

It has the explicit solution

$$V(t) = e^{-(\gamma-a-b)(t-t_0)}V_0 + \gamma e^{-(\gamma-a-b)t}\int_{t_0}^{t}\frac{e^{(\gamma-a-b)s}}{\widehat{N}(s)}\,ds.$$

The pullback limit as $t_0 \to -\infty$ exists only if $\gamma - a - b > 0$ and is then equal to

$$\widehat{V}(t) = \gamma e^{-(\gamma - a - b)t} \int_{-\infty}^t \frac{e^{(\gamma - a - b)s}}{\widehat{N}(s)} \, ds.$$

This gives a nontrivial nonautonomous equilibrium

$$\widehat{S}(t) = \widehat{N}(t) - \widehat{I}(t), \quad \widehat{I}(t) = \frac{e^{(\gamma - a - b)t}}{\gamma \int_{-\infty}^t \frac{e^{(\gamma - a - b)s}}{\widehat{N}(s)} \, ds}, \tag{6.9}$$

where

$$\widehat{N}(t) = a e^{-at} \int_{-\infty}^t q(s) e^{as} \, ds.$$

Remark 6.1. The $\widehat{S}(t)$, $\widehat{I}(t)$ and $\widehat{N}(t) = \widehat{S}(t) + \widehat{I}(t)$ defined here satisfy the SI equations (6.6)–(6.7).

Lemma 6.2. *The nonautonomous equilibrium solution $(\widehat{S}(t), \widehat{I}(t))$ is globally asymptotically stable in $\Sigma_2^{\pm}$ when $\gamma > a + b$.*

Proof. A general solution of the SI equations (6.6)–(6.7) will be compared with the nonautonomous equilibrium solution. From above

$$\left| N(t) - \widehat{N}(t) \right|^2 = e^{-2a(t-t_0)} \left| N(t_0) - \widehat{N}(t_0) \right| \to 0 \quad \text{as} \quad t \to \infty.$$

Now consider two solutions $I(t)$ and $\widehat{I}$ of the I equations with $I(0) > 0$, i.e.,

$$\frac{dI}{dt} = (\gamma - a - b)I - \gamma \frac{1}{N} I^2$$

and

$$\frac{d\widehat{I}}{dt} = (\gamma - a - b)\widehat{I} - \gamma \frac{1}{\widehat{N}} \widehat{I}^2,$$

respectively. These are Bernoulli equations which can be solved with the substitution $V = I^{-1}$, to give the linear ODE

$$\frac{dV}{dt} + (\gamma - a - b)V = \gamma \frac{1}{N}.$$

Writing $\widehat{V} = \widehat{I}^{-1}$, it follows that $\Delta V := V - \widehat{V}$ satisfies the linear ODE

$$\frac{d\Delta V}{dt} + \rho \Delta V = \gamma \left(\frac{1}{\widehat{N}} - \frac{1}{N} \right), \tag{6.10}$$

where

$$\rho := \gamma - a - b,$$

which is positive here.

This gives

$$\frac{d}{dt}(\Delta V)^2 + 2\rho(\Delta V)^2 = 2\gamma\left(\frac{1}{\widehat{N}} - \frac{1}{N}\right)\Delta V \leq 2\gamma\frac{|N - \widehat{N}|}{N\widehat{N}}|\Delta V|$$

$$\leq \frac{2\gamma}{(q^-)^2}|N - \widehat{N}||\Delta V| \leq \rho(\Delta V)^2 + \frac{4\gamma^2}{\rho(q^-)^4}|N - \widehat{N}|^2,$$

by Young's inequality, so

$$\frac{d}{dt}(\Delta V)^2 + \rho(\Delta V)^2 \leq \frac{4\gamma^2}{\rho(q^-)^4}|N(0) - \widehat{N}(0)|^2 e^{-2at}.$$

Integrating gives

$$\Delta V(t)^2 = (\Delta(0)^2 e^{-\rho t} + \frac{4\gamma^2}{\rho(q^-)^4}|N(0) - \widehat{N}(0)|^2 e^{-\rho t}\int_0^t e^{-2as}e^{\rho s}\,ds$$

$$\leq \Delta V(0)^2 e^{-\rho t} + \frac{4\gamma^2}{\rho(q^-)^4}|N(0) - \widehat{N}(0)|^2 e^{-\rho t}\int_0^t e^{(\rho-2a)s}\,ds$$

$$\leq \Delta V(0)^2 e^{-\rho t} + \frac{4\gamma^2}{\rho(\rho - 2a)(q^-)^4}|N(0) - \widehat{N}(0)|^2\left(e^{-\rho t} - e^{-2at}\right),$$

so

$$\Delta V(t) \to 0 \quad \text{as} \quad t \to 0.$$

(Note if $\rho = 2a$ the argument holds with a slight change in the use of Young's inequality.)

This says that

$$\frac{1}{I(t)} - \frac{1}{\widehat{I}(t)} \to 0 \quad \text{as} \quad t \to 0$$

and hence

$$I(t) - \widehat{I}(t) \to 0 \quad \text{as} \quad t \to 0$$

since $I(t) > 0$ and $\widehat{I}(t) > 0$ for all t. Now $|N(t) - \widehat{N}(t)| \to 0$ as $t \to 0$, so

$$\left|S(t) - \widehat{S}(t)\right| = \left|(N(t) - I(t)) - \left(\widehat{N}(t) - \widehat{I}(t)\right)\right|$$

$$\leq \left|N(t) - \widehat{N}(t)\right| + \left|I(t) - \widehat{I}(t)\right| \to 0 \quad \text{as} \quad t \to 0.$$

This completes the proof of Lemma 6.2. $\qquad\square$

6.3 The SI equations with variable interaction

Consider the SI equations for a constant limiting population $N = S + I \equiv 1$, but with a time-variable interaction term $\gamma = \gamma(t)$. It will be assumed that $\gamma \in C(\mathbb{R}, [\gamma^-, \gamma^+])$, where $0 < \gamma^- \leq \gamma^+ < \infty$. The equations now have the form

$$\frac{dS}{dt} = a - aS + bI - \gamma(t)SI, \tag{6.11}$$

$$\frac{dI}{dt} = -(a+b)I + \gamma(t)SI. \tag{6.12}$$

On adding the equations reduce to

$$\frac{dN}{dt} = a - aN,$$

which has the globally asymptotic solution $N(t) \equiv 1$, so the analysis can be restricted to the compact set

$$\Sigma_2 = \{(S, I) : S, I \geq 0, \ S + I \equiv 1\} \subset \mathbb{R}^2.$$

The equations (6.11)–(6.12) have an equilibrium solution $(\bar{S}, \bar{I}) = (1, 0)$ for all parameter values.

Lemma 6.3. *The nonautonomous equilibrium solution* $(\bar{S}, \bar{I}) = (1, 0)$ *is globally asymptotically stable in* Σ_2 *when* $\gamma^+ \leq a + b$. *It is unstable when* $\gamma^- > a + b$.

Proof. Suppose that $\gamma^+ < a + b$. Multiply the I equation by $2I$ to obtain

$$\frac{d}{dt}I^2 = -2(a+b)I^2 + 2\gamma(t)SI^2 \leq -2(a+b)I^2 + 2\gamma^+ I^2 \leq 2(\gamma^+ - a - b)I^2$$

because $0 \leq S \leq 1$ and $\gamma^+ > 0$. Hence

$$I^2(t) \leq I^2(0)e^{-2|\gamma^+ - a - b|t} \to 0 \quad \text{as} \quad t \to \infty.$$

The case $\gamma^+ = a + b$ is the same as in Lemma 6.1 with γ replaced by γ^+.

Now let $\gamma^- > a + b$ and suppose that $I \leq \varepsilon$ in the I equation, so $S \geq (1 - \varepsilon)$. Then

$$\frac{d}{dt}I = -(a+b)I + \gamma(t)SI \geq -(a+b)I + \gamma^-(1-\varepsilon)I \geq (\gamma^-(1-\varepsilon) - a - b)I,$$

so $I(t)$ is strictly increasing if $0 < I \leq \varepsilon N$, provided $\varepsilon < 1 - \frac{a+b}{\gamma^-}$. In particular, the equilibrium solution $(1, 0)$ is unstable when $\gamma^- > a + b$. $\square$

6.3.1 *Nontrivial limiting solutions*

Now consider the possibility of a nontrivial equilibrium, i.e., with $I \neq 0$, when $\gamma^- > a + b$. It cannot exist as a steady state, so a nonautonomous equilibrium will be sought.

The equation (6.12) for I can be rewritten

$$\frac{dI}{dt} = -(a+b)I + \gamma(t)I(1 - I),$$

that is as

$$\frac{dI}{dt} = (\gamma(t) - a - b)I - \gamma(t)I^2.$$

This is a Bernoulli equation which can be solved with the substitution $V = I^{-1}$, to give the linear ODE

$$\frac{dV}{dt} + (\gamma(t) - a - b)V = \gamma(t).$$

It has the explicit solution

$$V(t) = e^{-\int_{t_0}^{t} \rho(s)\, ds} V_0 + e^{-\int_{t_0}^{t} \rho(s)\, ds} \int_{t_0}^{t} \gamma(\tau) e^{\int_{t_0}^{\tau} \rho(s)\, ds}\, d\tau$$

$$= e^{-\int_{t_0}^{t} \rho(s)\, ds} V_0 + \int_{t_0}^{t} \gamma(\tau) e^{\int_{t_0}^{\tau} \rho(s)\, ds - \int_{t_0}^{t} \rho(s)\, ds}\, d\tau$$

$$= e^{-\int_{t_0}^{t} \rho(s)\, ds} V_0 + \int_{t_0}^{t} \gamma(\tau) e^{-\int_{\tau}^{t} \rho(s)\, ds}\, d\tau,$$

where $\rho(t) := \gamma(t) - a - b > 0$. Now

$$\int_{t_0}^{t} \rho(s)\, ds = \int_{t_0}^{t} (\gamma(s) - a - b)\, ds \geq (\gamma^- - a - b)(t - t_0) \geq 0.$$

The pullback limit as $t_0 \to -\infty$ exists only if $\gamma^- - a - b > 0$ and is then equal to

$$\widehat{V}(t) = \int_{-\infty}^{t} \gamma(\tau) e^{-\int_{\tau}^{t} \rho(s)\, ds}\, d\tau.$$

These integrals exists due to the assumptions on the coefficients. The nontrivial nonautonomous equilibrium

$$\widehat{S}(t) = 1 - \widehat{I}(t), \qquad \widehat{I}(t) = \left(\int_{-\infty}^{t} \gamma(\tau) e^{-\int_{\tau}^{t} \rho(s)\, ds}\, d\tau \right)^{-1}.$$

For γ a positive constant, they integrate out to give the nontrivial steady state solution

$$\bar{S} = \frac{a + b}{\gamma}, \qquad \bar{I} = \frac{\gamma - a - b}{\gamma}.$$

The proof of the following lemma is very similar to that of Lemma 6.2, but is given for completeness.

Lemma 6.4. *The nonautonomous equilibrium solution* $(\widehat{S}(t), \widehat{I}(t))$ *is globally asymptotically stable in* $\Sigma_2^{\pm}$ *when* $\gamma^- > a + b$.

Proof. Consider an arbitrary solution $I(t)$ of the I equation (6.12) with $I(0) > 0$ and the solution, i.e.,

$$\frac{dI}{dt} = (\gamma(t) - a - b)I - \gamma(t)I^2$$

and

$$\frac{d\widehat{I}}{dt} = (\gamma(t) - a - b)\widehat{I} - \gamma(t)\widehat{I}^2,$$

respectively. These are Bernoulli equations which can be solved with the substitution $V = I^{-1}$ to give the linear ODE

$$\frac{dV}{dt} + (\gamma(t) - a - b)V = \gamma(t).$$

Writing $\widehat{V} = \widehat{I}^{-1}$, it follows that $\Delta V := V - \widehat{V}$ satisfies the linear ODE

$$\frac{d\Delta V}{dt} + \rho(t)\Delta V = 0, \tag{6.13}$$

where $\rho(t) := \gamma(t) - a - b \geq \gamma^- - a - b > 0$. This gives

$$\Delta V(t)^2 = \Delta V(0)^2 e^{-2\int_0^t \rho(s)\,ds} \leq \Delta V(0)^2 e^{-2(\gamma^- - a - b)t},$$

so

$$\Delta V(t) \to 0 \quad \text{as} \quad t \to 0.$$

This says that

$$\frac{1}{I(t)} - \frac{1}{\widehat{I}(t)} \to 0 \quad \text{as} \quad t \to 0$$

and hence

$$I(t) - \widehat{I}(t) \to 0 \quad \text{as} \quad t \to 0$$

since $I(t) > 0$ and $\widehat{I}(t) > 0$ for all t. Similarly

$$\left| S(t) - \widehat{S}(t) \right| = \left| I(t) - \widehat{I}(t) \right| \to 0 \quad \text{as} \quad t \to 0.$$

This completes the proof of Lemma 6.4.   □

6.3.2   *Critical case*

The situation for $a + b \in (\gamma^-, \gamma^+]$ depends on how the function $\gamma(t)$ varies in time. Specifically, it will depend on the asymptotic behaviour of the integral

$$\frac{1}{t - t_0} \int_{t_0}^t \gamma(s)\,ds$$

as either $t \to \infty$ or $t_0 \to -\infty$. The limits usually will not exists, so the upper and lower limits should be used. These exist and lie between γ^- and γ^+.

In fact, here the disease free steady state $(1, 0)$ undergoes a nonautonomous bifurcation to the nonautonomous equilibrium solution $(\widehat{S}(t), \widehat{I}(t))$, see [Kloeden and Pötzsche (2015)].

6.4 The SIR model with variable population

The SIR equations with variable population(6.1)–(6.3) will be analysed here for the asymptotically stable limiting population $\widehat{N}(t)$ given by (6.5)

$$\frac{dS}{dt} = aq(t) - aS + bI - \gamma\frac{SI}{\widehat{N}(t)}, \tag{6.14}$$

$$\frac{dI}{dt} = -(a+b+c)I + \gamma\frac{SI}{\widehat{N}(t)}, \tag{6.15}$$

$$\frac{dR}{dt} = cI - aR. \tag{6.16}$$

These have a nonautonomous disease free equilibrium as for the SI equations for all parameter values. The dynamics beyond it becomes unstable and can be very complicated.

It is convenient here to eliminate the S variable and just consider the equations for the I and R variables in (6.14)–(6.16). Essentially, replace S by $S = \widehat{N} - I - R$ in the I equation (6.15) to obtain the IR system

$$\frac{dI}{dt} = (\gamma - a - b - c)I - \gamma\frac{I(I+R)}{\widehat{N}(t)}, \tag{6.17}$$

$$\frac{dR}{dt} = cI - aR, \tag{6.18}$$

where $0 \leq I, R \leq \widehat{N}(t)$, i.e., with solutions in time dependent triangular sets

$$\mathbb{T}_2(t) := \{(I,R) : I, R \geq 0, \ 0 \leq I + R \leq \widehat{N}(t)\},$$

hence in the larger common compact subset of $\mathbb{R}^2$

$$\mathbb{T}_2(t) \subset \{(I,R) : I, R \geq 0, \ 0 \leq I + R \leq g^+\}.$$

6.4.1 *Disease free nonautonomous equilibrium solution*

Note that $I(t) = R(t) \equiv 0$ is a solution. The S equation (6.14) then reduces to

$$\frac{dS}{dt} = aq(t) - aS$$

which is exactly the same as the equation for the total population, so it has pullback limiting

$$\widehat{S}_1(t) = \widehat{N}(t) = ae^{-at}\int_{-\infty}^{t} q(s)e^{as}\,ds.$$

The proof of the following lemma is very similar to that of Lemma 6.1, using reduced IR equations (6.17)–(6.18) for the stability proof and the original I equation (6.15) for the instability proof.

Lemma 6.5. *The disease free nonautonomous equilibrium solution $(\widehat{S}_1(t), 0, 0)$ is globally asymptotically stable in $\Sigma_3^{\pm}$ when $\gamma \leq a + b + c$. It is unstable when $\gamma > a + b - c$.*

6.4.2 *Nontrivial dynamics*

The situation is much more complicated for $\gamma > a + b - c$ when the disease free nonautonomous equilibrium solution is unstable. For this the reduced IR equations (6.17)–(6.18) will be used and will be rewritten as

$$\frac{dI}{dt} = \rho I - \gamma \frac{I(I+R)}{\widehat{N}(t)}, \qquad \frac{dR}{dt} = cI - aR, \tag{6.19}$$

where $\rho := \gamma - a - b - c > 0$. Since $R \geq 0$ the solution of the I equation in (6.19) satisfies the differential inequality

$$\frac{dI}{dt} \leq \rho I - \gamma \frac{I^2}{\widehat{N}(t)}.$$

This inequality written as an equality is a Bernoulli equation, which analogously with the derivation of the pullback solution (6.9) has the pullback limiting solution

$$\widehat{I}^+(t) = \frac{e^{\rho t}}{\gamma \int_{-\infty}^{t} \frac{e^{\rho s}}{\widehat{N}(s)} \, ds}, \qquad \widehat{N}(t) = ae^{-at} \int_{-\infty}^{t} q(s) e^{as} \, ds.$$

By a monotonicity argument, any pullback limiting solution of the reduced IR equations (6.19) thus has the upper bound

$$\widehat{I}(t) \leq \widehat{I}^+(t), \qquad \widehat{R}(t) \leq be^{-at} \int_{-\infty}^{t} e^{as} \widehat{I}^+(s) \, ds.$$

Such a pullback limiting solution arises as a nonautonomous bifurcation as ρ becomes positive. However, it need not be unique, especially for larger ρ. Indeed, [Stone, Shulgin and Agur (2001)] show that the dynamics can become chaotic when the population is constant and the interaction parameter $\gamma(t)$ is time periodic.

Remark 6.2. To proceed further with the analysis requires an appropriate concept of a nonautonomous attractor. The asymptotically stable nonautonomous equilibria constructed above by pullback convergence is an ideal candidate, but do not cover many of the situations that may arise.

Chapter 7

Attractors for processes

Let ϕ be a process on a complete metric space (X, d) with the time set $\mathbb{T}$.

We have seen that we should consider an invariant set of a nonautonomous dynamical system to be a family $\mathcal{A} = \{A_t, t \in \mathbb{T}\}$ of nonempty compact subsets A_t of X such that

$$\phi(t, t_0, A_{t_0}) = A_t \quad \text{for all } (t, t_0) \in \mathbb{T}_2^+.$$

Otherwise we lose some very interesting and useful behaviour.

In Example 5.4 we saw that the bounded entire solution of the linear ODE (5.3) forms a family of singleton sets

$$A_t = \left\{ \frac{1}{2}(\cos t + \sin t) \right\}, \quad t \in \mathbb{R},$$

to which all other solutions converge in the *pullback* sense i.e., as $t_0 \to -\infty$ with t held fixed, as well as in the *forward sense*, i.e., as $t \to \infty$ with t_0 held fixed. This is because $t - t_0 \to \infty$, so $e^{-(t-t_0)} \to 0$, in both cases. In Chapter 6 we saw other examples of such nonautonomous equilibria solutions.

In general, however, we can have pullback convergence without forward convergence or forward convergence without pullback convergence.

Example 7.1. The scalar nonautonomous ODE

$$\frac{dx}{dt} = 2tx \tag{7.1}$$

has the explicit solution $x(t, t_0, x_0) = x_0 e^{t^2 - t_0^2}$, for which

$$x_0 e^{t^2 - t_0^2} = x_0 e^{t^2} e^{-t_0^2} \to 0 \quad \text{as} \quad t_0 \to -\infty \qquad (\text{fixed } t).$$

Here there is a pullback attractor with singleton sets $A_t = \{0\}$ for all $t \in \mathbb{R}$. There is no forward attraction here.

Example 7.2. The scalar nonautonomous ODE

$$\frac{dx}{dt} = -2tx$$

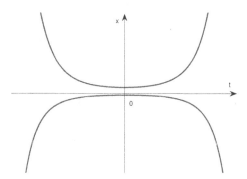

Fig. 7.1   Typical solutions of the nonautonomous ODE (7.1).

has the explicit solution $x(t, t_0, x_0) = x_0 e^{-t^2 + t_0^2}$, for which

$$x_0 e^{-t^2 + t_0^2} = x_0 e^{t_0^2} e^{-t^2} \to 0 \quad \text{as} \quad t \to \infty \qquad \text{(fixed } t_0\text{)}.$$

Here there is a forward attractor with singleton component sets $A_t = \{0\}$ for all $t \in \mathbb{R}$. There is no pullback attraction here.

Thus we have *two* types of nonautonomous attractors for processes, one with pullback convergence and one with forward convergence.

Definition 7.1. A family $\mathcal{A} = \{A_t, t \in \mathbb{T}\}$ of nonempty compact subsets of X, which is ϕ-invariant, is called a

- *pullback attractor* if it pullback attracts all bounded subsets D of X, i.e.,

$$\lim_{t_0 \to -\infty} \mathrm{dist}_X \left(\phi(t, t_0, D), A_t \right) = 0, \qquad \text{(fixed } t\text{)}$$

- *forward attractor* if it forward attracts all bounded subsets D of X, i.e.,

$$\lim_{t \to \infty} \mathrm{dist}_X \left(\phi(t, t_0, D), A_t \right) = 0, \qquad \text{(fixed } t_0\text{)}.$$

Remark 7.1. To enable us to handle *non-uniformities*, which are common in nonautonomous systems, we often replace a "bounded set" D by a "family of appropriately chosen bounded sets" $\mathcal{D} = \{D_t, t \in \mathbb{T}\}$. The limits in the above definition are then for the expression

$$\mathrm{dist}_X \left(\phi\left(t, t_0, D_{t_0}\right), A_t \right).$$

Note that the bounded sets in the family $\mathcal{D} = \{D_t, t \in \mathbb{T}\}$ may have to be restricted in some way to ensure that the absorbing property holds in given examples. For example, they could be *uniformly bounded*, i.e., if there is a common bounded subset B of X such that $D_t \subseteq B$ for all $t \in \mathbb{T}$. More generally, the subsets D_t could be allowed to grow but not too quickly as $t \to \pm\infty$. See Definition 9.4 of a *tempered* family of bounded subsets and the discussion in Section 9.3 in the context of skew product flows.

Similarly, we say that a pullback attractor $\mathcal{A} = \{A_t, t \in \mathbb{T}\}$ is *uniformly bounded* if $\bigcup_{t \in \mathbb{T}} A_t$ is bounded or, equivalently, if there is a common bounded subset B of X such that $A_t \subseteq B$ for all $t \in \mathbb{T}$.

We have the following characterisation of a uniformly bounded pullback attractor.

Proposition 7.1. *A uniformly bounded pullback attractor* $\mathcal{A} = \{A_t, t \in \mathbb{T}\}$ *of a process* ϕ *is uniquely determined by the bounded entire solutions of the process, i.e.,*

$$x_0 \in A_{t_0} \quad \Longleftrightarrow \quad \exists \text{ a bounded entire solution } e(\cdot) \text{ with } e(t_0) = x_0.$$

Proof. *(Sufficiency)* Pick $t_0 \in \mathbb{T}$ and $x_0 \in A_{t_0}$ arbitrarily. Then, due to the ϕ-invariance of the pullback attractor there exists an entire solution $e(\cdot)$ with $e(t_0) = x_0$ and $e(t) \in A_t$ for all $t \in \mathbb{T}$. (See Lemma 4.1 in Chapter 4). Moreover,

$$e(t) \in A_t \subseteq B$$

for all $t \in \mathbb{T}$, so the entire solution $e(\cdot)$ is a uniformly bounded entire solution.

(Necessity) Let $e(\cdot)$ be a bounded entire solution of the process ϕ. Then the set $D_e := \{e(t), t \in \mathbb{T}\}$ is a bounded subset of X. Since $\mathcal{A}$ pullback attracts bounded subsets of X, for each $t \in \mathbb{T}$ we have

$$0 \leq \mathrm{dist}_X \left(e(t), A_t \right) \leq \mathrm{dist}_X \left(D_e, A_t \right) \to 0 \quad \text{as } t_0 \to -\infty,$$

from which we conclude that

$$\mathrm{dist}_X \left(e(t), A_t \right) = 0 \qquad \text{for each } t \in \mathbb{T},$$

i.e., $e(t) \in A_t$ for every $t \in \mathbb{T}$ since the A_t are compact sets. $\qquad \square$

7.1 Existence of a pullback attractor

In analogy to the autonomous case we can show that the existence of a pullback absorbing compact set implies the existence of a pullback attractor. Similarly, the compactness of the absorbing set can be replaced by closed and boundedness provided the process satisfies some additional compactness property.

To handle non-uniformities in the dynamics we also consider a pullback absorbing family of sets instead of a single set and assume that it absorbs a family of nonempty bounded subsets, which will be assumed to be uniformly bounded here (but this can be weakened as indicated in Remark 7.1).

Definition 7.2. A family $\mathcal{B} = \{B_t, t \in \mathbb{T}\}$ of nonempty subsets of X is called a *pullback absorbing family* process ϕ on X if for each $t \in \mathbb{T}$ and every uniformly bounded family $\mathcal{D} = \{D_t, t \in \mathbb{T}\}$ nonempty bounded subsets of X there exists a $T_{t,\mathcal{D}} \in \mathbb{T}^+$ such that

$$\phi\left(t, t_0, D_{t_0}\right) \subseteq B_t \qquad \text{for all } t_0 \leq t - T_{t,\mathcal{D}}.$$

Definition 7.3. A family $\mathcal{B} = \{B_t, t \in \mathbb{T}\}$ of nonempty subsets of X is called *positive invariant* with respect to a process ϕ or *ϕ-positive invariant* if

$$\phi(t, t_0, B_{t_0}) \subseteq B_t \qquad \text{for all } t \geq t_0.$$

Theorem 7.1 (Existence of a pullback attractor). *Suppose that a process ϕ on a complete metric space (X, d) has a ϕ-positive invariant pullback absorbing family $\mathcal{B} = \{B_t, t \in \mathbb{T}\}$ of compact sets.*

Then ϕ has a global pullback attractor $\mathcal{A} = \{A_t, t \in \mathbb{T}\}$ with component subsets determined by

$$A_t = \bigcap_{t_0 \leq t} \phi(t, t_0, B_{t_0}) \qquad \text{for each } t \in \mathbb{T}. \tag{7.2}$$

Moreover, if $\mathcal{A}$ is uniformly bounded then it is unique.

Proof The set A_t is nonempty and compact as the nested intersection of compact subsets since

$$\phi(t, t_0', B_{t_0}) = \phi\left(t, t_0, \phi\left(t_0, t_0', B_{t_0'}\right)\right) \subseteq \phi(t, t_0, B_{t_0}), \qquad t_0' \leq t_0.$$

Let $\mathcal{B} = \{B_t, t \in \mathbb{T}\}$ be the pullback absorbing family and let A_t be defined by (7.2). Clearly,

$$A_t \subseteq B_t \qquad \text{for each } t \in \mathbb{T}.$$

In fact, $A_t \subseteq \phi(t, t_0, B_{t_0}) \subseteq B_t$ for any $t_0 \leq t$.

Step 1. First, we show for any $t \in \mathbb{T}$ that

$$\lim_{t_0 \to -\infty} \text{dist}_X\left(\phi(t, t_0, B_{t_0}), A_t\right) = 0.$$

Suppose not. Then there exist sequences $k \to \infty$ and

$$x_k \in \phi(t, t_k, B_{t_k}) \subseteq B_t \qquad \text{with} \quad t_k \to -\infty$$

such that

$$\text{dist}_X(x_k, A_t) \geq \varepsilon_0, \qquad \text{for all} \quad k,$$

for some $\varepsilon_0 > 0$.

The subset $\{x_k\}_{k \in \mathbb{N}}$ of B_t is relatively compact, so there is a convergent subsequence $k_j \to \infty$ and a point $x_0 \in B_t$ such that $x_{k_j} \to x_0$ as $k_j \to \infty$. Now

$$x_{k_j} \in \phi\left(t, t_{k_j}, B_{t_{k_j}}\right) \subseteq \phi(t, t_k, B_{t_k})$$

for all all $k_j \geq k$ and each k. This implies that

$$x_0 \in \phi(t, t_k, B_{t_k}) \qquad \text{for all} \quad k \in \mathbb{N},$$

so

$$x_0 \in \bigcap_{k \geq 0} \phi\left(t, t_k, B_{t_k}\right),$$

hence $x_0 \in A_t$, which is a contradiction. Hence the assertion of Step 1 has been proved.

Step 2. From the above convergence, for every $\varepsilon > 0$ and $t \in \mathbb{T}$ there exists $T_{\varepsilon,t} \geq 0$ such that

$$\text{dist}_X\left(\phi\left(t, t_0, B_{t_0}\right), A_t\right) < \varepsilon$$

for all $t_0 \leq t - T_{\varepsilon,t}$.

Let D be any bounded subset of X. By the pullback absorption property of $\mathcal{B} = \{B_t, t \in \mathbb{T}\}$ we have

$$\phi\left(t, t_0, D\right) \subseteq B_t$$

for all $t_0 < t$ with $t - t_0$ large enough. Let $t_0'' < t_0' < t$. Then by the 2-parameter semi-group property

$$\phi\left(t, t_0'', D\right) = \phi\left(t, t_0', \phi\left(t_0', t_0'', D\right)\right) \subseteq \phi\left(t, t_0', B_{t_0'}\right)$$

provided $t_0' - t_0''$ is large enough (for the last inclusion).

Step 3. The ϕ-invariance of the family $\mathcal{A} = \{A_t, t \in \mathbb{T}\}$ will now be shown.

By (7.2) the set $F_{t_0}(t) := \phi\left(t, t_0, B_{t_0}\right) \subseteq B_t$ for all $t_0 \leq t$ and by definition

$$A_t = \bigcap_{t_0 < t} F_{t_0}(t).$$

Similarly, $A_\tau = \bigcap_{t_0 < \tau} F_{t_0}(\tau)$ for any $\tau < t$. Clearly

$$\phi\left(t, \tau, \bigcap_{t_0 < \tau} F_{t_0}(\tau)\right) \subset \bigcap_{t_0 < \tau} \phi\left(t, \tau, F_{t_0}(\tau)\right).$$

To show the $\supseteq$ inclusion, let

$$x \in \bigcap_{t_0 < \tau} \phi\left(t, \tau, F_{t_0}(\tau)\right).$$

Then there exist $x_m \in F_{t_m}(\tau) \subseteq B_\tau$ such that $x = \phi\left(t, \tau, x_m\right)$.

Now the sets $F_{t_m}(\tau)$ are compact and monotonically decreasing with increasing m (i.e., with $t_m \to -\infty$), so the set $\{x_m\}$ has a limit point $\hat{x} \in \bigcap_{t_m < \tau} F_{t_m}(\tau)$. By continuity of $\phi(t, t_0, \cdot)$ it follows that $x = \phi(t, t_0, \hat{x})$. Thus

$$x \in \phi\left(t, \tau, \bigcap_{t_m < \tau} F_{t_m}(\tau)\right) = \phi\left(t, \tau, A_\tau\right).$$

This completes the proof that

$$\bigcap_{t_m < \tau} \phi\left(t, \tau, F_{t_m}(\tau)\right) \subseteq \phi\left(t, \tau, \bigcap_{t_m < \tau} F_{t_m}(\tau)\right).$$

By this, the compactness of $F_{t_m}(\tau)$ and the continuity of $\phi(t,\tau,\cdot)$ we have

$$\phi(t,\tau,A_\tau) = \bigcap_{t_m<\tau} \phi(t,\tau,F_{t_m}(\tau)) = \bigcap_{t_m<\tau} \phi(t,\tau,\phi(\tau,t_m,B_{t_m}))$$

$$= \bigcap_{t_m<\tau} \phi(t,t_m,B_{t_m}) \qquad \text{two-parameter semi-group property}$$

$$\supseteq \bigcap_{t_m<t} \phi(t,t_m,B_{t_m}) \supseteq A_t,$$

i.e., $A_t \subseteq \phi(t,\tau,A_\tau)$ for all $t \geq \tau$.

Relabelling, we have

$$A_\tau \subseteq \phi(\tau,t_m,A_{t_m}) \qquad \text{for all } \tau \geq t_m.$$

Thus

$$\phi(t,\tau,A_\tau) \subseteq \phi(t,\tau,\phi(\tau,t_m,A_{t_m})) \qquad \text{two-parameter semi-group property}$$

$$\subseteq \phi(t,\tau,\phi(\tau,t_m,B_{t_m})) \qquad \text{(since } A_{t_m} \subseteq B_{t_m})$$

$$= \phi(t,t_m,B_{t_m}) \qquad \text{two-parameter semi-group property}$$

$$\subseteq B_\varepsilon(A_t)$$

for all t_m with $t - t_m$ large enough (depending on $\varepsilon > 0$). With $t_m \to -\infty$ or equivalently $\varepsilon \to 0$ we have

$$\phi(t,\tau,A_\tau) \subseteq A_t.$$

This completes the proof that $\mathcal{A} = \{A_t, t \in \mathbb{T}\}$ is ϕ-invariant and hence that it is a pullback attractor.

Step 4.   Suppose that $\mathcal{A}$ is uniformly bounded. Then, see Proposition 7.1, it is characterised by the set of bounded entire solutions of ϕ, which is unique. Hence $\mathcal{A}$ is also unique.

This completes the proof of Theorem 7.1 on the existence of a pullback attractor for a process.   □

Remark 7.2. The assumption that the pullback absorbing family $\mathcal{B} = \{B_t, t \in \mathbb{T}\}$ is positive invariant simplifies the proof, but is not a big restriction. As we saw in the autonomous case (see Chapter 5), if it is not positively invariant, then we can replace it by a bigger set which is positively invariant. Such the family $\mathcal{B}^* = \{B_t^*, t \in \mathbb{T}\}$ has component sets B_t^* defined by

$$B_t^* := \bigcup_{t-T_{\mathcal{B},t} \leq \tau \leq t} \phi(t,\tau,B_\tau),$$

where $T_{\mathcal{B},t}$ is the absorbing time for $\mathcal{B}$ to pullback itself into B_t.

Remark 7.3. Theorem 7.1 characterises and gives the existence of a pullback attractor. Although forward attractors can be constructed by a similar pullback argument within the absorbing set (see Theorem 11.2), this provides only a necessary condition for the family of sets so obtained to be a forward attractor. Moreover, forward attractors need not be unique. For example, in Example 7.2, the family of singleton sets $A_t = \{0\}$ is a forward attractor, but so too is the family of sets $A_t^{(r)} = re^{-t^2+t_0^2}[-1,1]$ for each $r > 0$. This situation is discussed in Chapter 5.

Remark 7.4. When the absorbing sets are compact the expression (7.2) for the pullback attractor components sets in Theorem 7.1 is the intersection of nested compact sets and thus nonempty and compact. Such absorbing sets are not difficult to determine in locally compact spaces such as $\mathbb{R}^d$ since compactness is equivalent to closedness and boundedness.

This equivalence does not hold in most infinite dimensional spaces, such as the state spaces of many evolution equations. Since compact absorbing sets are much more difficult to find than closed and bounded absorbing sets in such spaces, the latter are used with an additional property of the compactness or asymptotic compactness of the process ϕ. The compactness of ϕ means that ϕ maps closed and bounded subsets to precompact sets. It is a strong property. Asymptotic compactness is more general. It has two versions depending on if it is defined in terms of forward or pullback convergence. See, e.g., [Carvalho, Langa and Robinson (2013); Chepyzhov and Vishik (2002); Hale (1988); Ladyzhenskaya (1991)].

Definition 7.4. A process ϕ on a Banach space X is said to be pullback asymptotically compact if, for each $t \in \mathbb{R}$, each sequence $\{s_k\}_{k\in\mathbb{N}}$ in $\mathbb{R}$ with $s_k \leq t$ and $s_k \to -\infty$ as $k \to \infty$, and each bounded sequence $\{x_k\}_{k\in\mathbb{N}}$ in X, the sequence $\{\phi(t, s_k, x_k)\}_{k\in\mathbb{N}}$ has a convergent subsequence.

7.2 Strictly contracting processes: an alternative existence proof

Theorem 7.1 on the existence of a pullback attractor for a process is mainly useful in locally compact $\mathbb{R}^d$ state spaces where compact subsets are equivalently closed and bounded. Compact subsets are not as common in infinite dimensional state spaces so absorbing sets are often assumed just to be closed and bounded. Then some additional compactness or asymptotic compactness property of the process is required, e.g., $\phi(t, t_0, B)$ is compact for all bounded sets B and $t > t_0$.

In the special case that the process satisfies a uniform contractivity condition as in Example 5.4, a Cauchy sequence argument can be used instead of compactness to obtain the existence of a pullback attractor.

Definition 7.5. A process ϕ on a complete metric space (X, d_X) is said to satisfy a uniformly strictly contracting property, i.e., there is an $L > 0$ such that

$$d_X\left(\phi(t, t_0, x_0), \phi(t, t_0, y_0)\right) \leq e^{-L(t-t_0)} \cdot d_X\left(x_0, y_0\right), \qquad (7.3)$$

for all x_0, $y_0 \in X$, $(t, t_0) \in \mathbb{R}^2_\geq$.

In this case the pullback absorbing family need only consist of closed and bounded subsets of a complete metric space (X, d_X), thus not necessarily compact.

Theorem 7.2. *Suppose that a process ϕ on complete metric space (X, d_X) satisfies a uniformly strictly contracting property and has a positively invariant pullback absorbing family $\mathcal{B} = \{B_t : t \in \mathbb{R}\}$ of nonempty closed and uniformly bounded subsets of X, i.e., there is a finite $R > 0$ such that $\|B_t\| := \sup_{b \in B_t} d_X(b, 0) \leq R$ for all $t \in \mathbb{R}$.*

Then ϕ has a pullback attractor $\mathcal{A} = \{A_t\}_{t \in \mathbb{R}}$ with component sets consisting of singleton sets, i.e., $A_t = \{\chi^*(t)\}$ for each $t \in \mathbb{R}$, where $\chi^*(\cdot)$ is an entire solution of the process, i.e., with $\chi^*(t) = \phi(t, t_0, \chi^*(t_0))$ for all $(t, t_0) \in \mathbb{R}^2_\geq$.

Moreover, $\mathcal{A} = \{\chi^*(t)\}$ is also forward attracting.

Proof. *(Existence)* Fix any monotone sequence $\{t_n\}_{n \in \mathbb{N}}$ of initial times in $(-\infty, 0)$ with $t_n \to -\infty$ as $n \to \infty$. Furthermore, select arbitrarily, $x_n \in B_{t_n}$ for each $n \in \mathbb{N}$.

At time $t = 0$, consider the sequence $\{\varphi_n\}$ defined by

$$\varphi_n := \phi(0, t_n, x_n). \tag{7.4}$$

Then φ_n belongs to B_0 for each $n \in \mathbb{N}$. By Lemma 7.1 below it is a Cauchy sequence, which thus has a unique limit x_0^* in B_0.

Now define $\chi^*(t) := \phi(t, 0, x_0^*)$ for $t \geq 0$ and repeat the above argument with $t = 0$ replaced by -1, i.e., consider the sequence $\varphi_n^{(-1)} := \phi(-1, t_n, x_n)$ in B_{-1} (for the same $x_n \in B_{t_n}$ and the same sequence $t_n \to -\infty$) to obtain a limit $x_{-1}^* \in B_{-1}$ such that $x_0^* = \phi(0, -1, x_{-1}^*)$. In addition, define $\chi^*(t) = \phi(t, -1, x_{-1}^*)$ for $-1 < t < 0$.

Then proceed with this construction via induction for each $-n$ and $-n - 1$ to obtain a limit $x_{-n-1}^* \in B_{-n-1}$ with $x_{-n}^* = \phi(-n, -n - 1, x_{-n-1}^*)$ and define $\chi^*(t) = \phi(t, -n - 1, x_{-n-1}^*)$ for $-n - 1 < t < -n$. In this way an entire solution χ^* of the process is constructed, i.e., with $\chi^*(t) = \phi(t, s, \chi^*(s))$ for all $(t, s) \in \mathbb{R}^2_\geq$.

Hence, by the uniform strictly contracting condition (7.3), for any t_0 and $x_0 \in B_{t_0}$,

$$d_X(\phi(t, t_0, x_0), \chi^*(t)) \leq e^{-L(t-t_0)} \cdot d_X(x_0, \chi^*(t_0)) \leq 2Re^{-L(t-t_0)},$$

since

$$d_X(x_0, \chi^*(t_0)) \leq d_X(x_0, 0) + d_X(\chi^*(t_0), 0) \leq \|B_{t_0}\| + \|B_{t_0}\| = 2\|B_{t_0}\| \leq 2R.$$

Hence all solutions of the process are pullback attracted by $\chi^*(t)$.

(Uniqueness) It follows by a contradiction argument as in the proof of Proposition 8.1 that any other entire solution taking values in the sets B_t must coincide with $\chi^*(\cdot)$. Indeed, let $\bar{\chi}_t^*$ be another such entire solution and suppose that $\|\chi^*(0) - \bar{\chi}^*(0)\| \geq \varepsilon_0 > 0$.

Similarly to the proof of Lemma 7.1 below, the pullback contracting condition (7.3) gives

$$d_X\left(\phi(0,s,\chi^*(s)),\phi(0,s,\bar{\chi}^*(s))\right) \leq e^{Ls} \cdot d_X\left(\chi^*(s),\bar{\chi}^*(s)\right) \leq 2Re^{Ls}$$

for all $s \leq 0$ since

$$d_X\left(\chi^*(s),\bar{\chi}^*(s)\right) \leq d_X\left(\chi^*(s),0\right) + d_X\left(\bar{\chi}^*(s),0\right) \leq 2R.$$

Defining $T < 0$ such that $2Re^{Ls} \leq \varepsilon_0/2$ for all $s \leq T$, gives

$$d_X\left(\phi(0,s,\chi^*(s)),\phi(0,s,\bar{\chi}^*(s))\right) \leq \frac{1}{2}\varepsilon_0$$

for all $s \leq T$. However, $\bar{\chi}^*(0) = \phi(0,s,\bar{\chi}^*(s))$ and $\chi^*(0) = \phi(0,s,\chi^*(s))$, so

$$\varepsilon_0 \leq d_X\left(\chi^*(0),\bar{\chi}^*(0)\right) = d_X\left(\phi(0,s,\chi^*(s)),\phi(0,s,\bar{\chi}^*(s))\right) \leq \frac{1}{2}\varepsilon_0$$

for all $s \leq T$, which is a contradiction.

Thus the process ϕ has a pullback attractor $\mathcal{A} = \{A_t, t \in \mathbb{R}\}$ with component sets $A_t = \{\chi^*(t)\}$ for every $t \in \mathbb{R}$.

(Forward attraction) By the uniform contraction property (7.3) and the uniform boundedness of the component subsets of the absorbing family

$$d_X\left(\phi(t,t_0,x_0),\chi^*(t)\right) \leq e^{-L(t-t_0)} \cdot d_X\left(x_0,\chi^*(t_0)\right) \leq 2Re^{-L(t-t_0)} \to 0 \text{ as } t \to \infty,$$

so the family $\mathcal{A} = \{\{\chi^*(t)\}, t \in\in \mathbb{R}\}$ is also forward attracting.

This completes the proof of Theorem 7.2. $\qquad\qquad\square$

A key step in the proof of Theorem 7.2 uses the following lemma.

Lemma 7.1. *Under the assumptions of Theorem 7.2, the sequence $\{\varphi_n\}$ in B_0 given by (7.4) is a Cauchy sequence in X with values in B_0. In particular, there exists a unique limit $x_0^* \in B_0$ such that*

$$d_X\left(\varphi_n,x_0^*\right) \longrightarrow 0 \quad \text{as } n \to \infty.$$

Proof. It needs to be shown that for every $\varepsilon > 0$ there exists an integer $N_\varepsilon > 0$ with

$$d_X\left(\varphi_n,\varphi_m\right) \leq \varepsilon \quad \text{for all } n, m \geq N_\varepsilon.$$

Then $\{\varphi_n\}$ is a Cauchy sequence in X for the norm $\|\cdot\|$ and the completeness of X provides the unique limit $x_0^* \in X$.

Let $\{t_n\}_{n\in\mathbb{N}}$ be some monotone decreasing sequence tending to $-\infty$ as in the proof of Theorem 7.2. Choose the indices $m > n$ arbitrarily. Then by the 2-parameter semigroup property

$$\varphi_m \stackrel{\text{Def.}}{=} \phi(0,t_m,x_m) = \phi\left(0,t_n,\phi(t_n,t_m,x_m)\right) = \phi\left(0,t_n,\hat{x}_{n,m}\right),$$

where $\hat{x}_{n,m} := \phi(t_n, t_m, x_m) \in B_{t_n}$ due to the positive invariance of the pullback absorbing family and the construction of the time sequence. By the uniform strict contracting condition (7.3) it thus follows that

$$d_X(\varphi_n, \varphi_m) = d_X\left(\phi(0, t_n, x_n), \phi(0, t_n, \hat{x}_{n,m})\right)$$

$$\leq e^{Lt_n} \cdot d_X(x_n, \hat{x}_{m,n}) \leq 2\, e^{Lt_n}\, \|B_{t_n}\|$$

since

$$d_X(x_n, \hat{x}_{m,n}) \leq d_X(x_n, 0) + d_X(\hat{x}_{m,n}, 0) \leq 2\|B_{t_n}\| \leq 2R.$$

Hence $\{\varphi_n\}$ is a Cauchy sequence with N_ε chosen as the smallest integer $n \in \mathbb{N}$ for which $2Re^{Lt_n} = 2Re^{-L|t_n|} \leq \varepsilon$, since $t_n \to -\infty$ as $n \to \infty$. $\qquad\square$

Chapter 8

Examples of pullback attractors for processes

The pullback attractor can be determined explicitly in simple examples, but in more complicated examples all one can do is to show that a pullback attractor exists in a known absorbing set.

8.1 A linear differential equation

Consider again Example 5.4 with the scalar linear ODE

$$\frac{dx}{dt} + x = \cos t$$

with the explicit solution

$$x(t) = x_0 e^{-(t-t_0)} - \frac{1}{2} \left(\cos t_0 + \sin t_0 \right) e^{-(t-t_0)} + \frac{1}{2} \left(\cos t + \sin t \right).$$

It has a pullback attractor consisting of a single entire solution and singleton component sets

$$A_t = \left\{ \frac{1}{2} (\cos t + \sin t) \right\}, \qquad t \in \mathbb{R}.$$

The pullback attractor is also a forward attractor since the difference of any two solutions satisfies

$$\frac{d}{dt} \left(x(t) - y(t) \right) = - \left(x(t) - y(t) \right)$$

which implies that

$$|x(t) - y(t)| \le |x_0 - y_0| \, e^{-(t-t_0)}, \tag{8.1}$$

so the system is uniformly strictly contractive.

This implies, amongst other things, that all solutions converge together forward in time and, in particular, that all solutions converge forward in time to the solution $\bar{x}(t)$, i.e.,

$$\lim_{t \to \infty} |x(t, t_0, x_0) - \bar{x}(t)| = 0, \quad (t_0 \text{ held fixed}).$$

8.2   A linear difference equation

Consider the linear difference equation

$$x_{n+1} = ax_n + b_n, \tag{8.2}$$

where $|a| < 1$ and $|b_n| \leq B < \infty$. Then

$$x_{n_0+1} = ax_{n_0} + b_{n_0},$$

$$x_{n_0+2} = ax_{n_0+1} + b_{n_0+1} = a\left[ax_{n_0} + b_{n_0}\right] + b_{n_0+1} = a^2 x_{n_0} + ab_{n_0} + b_{n_0+1},$$

$$x_{n_0+3} = ax_{n_0+2} + b_{n_0+2} = a\left[a^2 x_{n_0} + ab_{n_0} + b_{n_0+1}\right] + b_{n_0+2}$$

$$= a^3 x_{n_0} + a^2 b_{n_0} + ab_{n_0+1} + b_{n_0+2}.$$

In general,

$$x_n = a^{n-n_0} x_0 + \sum_{j=0}^{n-n_0-1} a^{n-n_0-1-j} b_{n_0+j}$$

or (better for taking the pullback limit)

$$x_n = a^{n-n_0} x_0 + \sum_{j=n_0}^{n-1} a^{n-1-j} b_j.$$

The pullback limit as $n_0 \to -\infty$ gives the entire sequence

$$\bar{x}_n = \sum_{j=-\infty}^{n-1} a^{n-1-j} b_j, \qquad n \in \mathbb{Z}.$$

This is indeed a convergent series and is uniformly bounded by

$$|\bar{x}_n| \leq \sum_{j=-\infty}^{n-1} |a|^{n-1-j} |b_j| \leq B \sum_{j=-\infty}^{n-1} |a|^{n-1-j}$$

$$= B \sum_{k=0}^{\infty} |a|^k = \frac{B}{1 - |a|} < \infty.$$

Exercise 8.1. As an exercisShow that $\bar{x}_{n+1} = a\bar{x}_n + b_n$, i.e., the bi-infinite sequence $\{\bar{x}_n, n \in \mathbb{Z}\}$ is an entire solution of the linear difference equation (8.2).

The pullback attractor here also consists of singleton component sets (i.e., a single entire solution)

$$A_n = \{\bar{x}_n\}, \qquad n \in \mathbb{Z}.$$

It is also a forward attractor because the difference of any two solutions satisfies

$$|x_n - y_n| = a^{n-n_0} |x_{n_0} - y_{n_0}|.$$

The linear difference equation (8.2) thus satisfies the uniform strict contractivity property.

8.3 A Nonlinear differential equation

Now consider a nonautonomous Bernoulli differential equation

$$\frac{dx}{dt} = ax - b(t)x^3, \tag{8.3}$$

where a is a constant and $b : \mathbb{R} \to \mathbb{R}$ is continuous and satisfies $b(t) \in [b_0, b_1]$ for all $t \in \mathbb{R}$ where $0 < b_0 < b_1 < \infty$.

Assuming $b(t)$ is nonconstant, the nonautonomous differential equation has only one equilbrium point 0 and this exists for all values of a. Moreover, when $a \leq 0$, using the Lyapunov function $V(x) = x^2$ gives

$$\frac{d}{dt}V(x_a(t)) = 2ax_a(t)^2 - 2b(t)x_a(t)^4$$

$$\leq 2aV(x_a(t)) - 2b_0 V(x_a(t))^2$$

$$\leq -2b_0 V(x_a(t))^2.$$

Hence

$$V(x_a(t)) \leq \frac{V(x_a(t_0))}{1 + 2b_0\ V(x_a(t_0))\ (t - t_0)} \to 0 \quad \text{as } t \to \infty,$$

where $x_a(t)$ is any solution of the ODE (8.3) for $a \leq 0$.

The equilibrium point 0 loses stability in the linearized ODE at $a = 0$ and is unstable for $a > 0$. Unlike the autonomous version of the ODE (8.3) no new equilibrium points come into existence with a pitchfork bifurcation when $a > 0$. Instead there are two bounded nonautonomous equilbria solutions $\pm\bar{\phi}_a$, where

$$\bar{\phi}_a(t) = \frac{1}{\sqrt{2\int_{-\infty}^{t} b(s)e^{-2a(t-s)}\,ds}}, \quad t \in \mathbb{R}, \tag{8.4}$$

which satisfies $\bar{\phi}_{\dot{a}}(t) \in \left[\sqrt{a/b_1}, \sqrt{a/b_0}\right]$ for all $t \in \mathbb{R}$. Moreover, each of the solutions $\pm\bar{\phi}_a$ is locally asymptotically stable in $\mathbb{R}^+$ and $\mathbb{R}^-$, respectively.

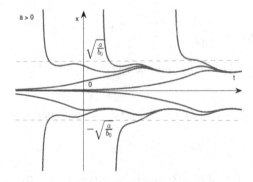

Fig. 8.1   Typical solutions of the nonautonomous ODE (8.3) for $a > 0$.

The proof follows from the fact that the Bernoulli equation (8.3) can be transformed into the linear differential equation

$$\frac{dv}{dt} + 2av = 2b(t)$$

with the substitution $v = x^{-2}$ (recall that $x = 0$ is an equilibrium of the Bernoulli equation (8.3)). This linear differential equation can be integrated to give

$$\frac{1}{x_a(t, t_0, x_0)^2} = \frac{1}{x_0^2} e^{-2a(t-t_0)} + 2 \int_{t_0}^{t} b(s) e^{-2a(t-s)} \, ds. \tag{8.5}$$

Holding t and x_0 fixed in (8.5) and taking the pullback limit as $t_0 \to -\infty$ gives the pullback limit solution

$$\bar{\phi}_a(t) = \lim_{t_0 \to -\infty} x_a(t, t_0, x_0)$$

given by (8.4). This is itself an entire solution of the differential equation (8.3) and thus satisfies (8.5) with $x_0 = \bar{\phi}_a(t_0)$. Specifically

$$\frac{1}{\bar{\phi}_a(t)^2} = \frac{1}{\bar{\phi}_a(t_0)^2} e^{-2a(t-t_0)} + 2 \int_{t_0}^{t} b(s) e^{-2a(t-s)} \, ds. \tag{8.6}$$

To show that $\bar{\phi}_a(t)$ is asymptotically stable for all $x_0 > 0$ subtract (8.5) from (8.6) to obtain

$$\frac{1}{\bar{\phi}_a(t)^2} - \frac{1}{x_a(t, t_0, x_0)^2} = \left(\frac{1}{\bar{\phi}_a(t_0)^2} - \frac{1}{x_0^2} \right) e^{-2a(t-t_0)} \to 0 \quad \text{as } t \to \infty,$$

from which the result follows.

In fact, $\bar{\phi}_a(t)$ is uniformly asymptotically Lyapunov stable in $\mathbb{R}^+$, since $\bar{\phi}_a(t) \in \left[\sqrt{a/b_1}, \sqrt{a/b_0} \right]$, so for every $\epsilon > 0$ and $x_0 > 0$ there exists $T(x_0, \epsilon) \geq 0$ such that

$$\left(\frac{1}{\bar{\phi}_a(t_0)^2} - \frac{1}{x_0^2} \right) e^{-2a(t-t_0)} \leq \left(\frac{b_1}{a} - \frac{1}{x_0^2} \right) e^{-2a(t-t_0)} < \epsilon \tag{8.7}$$

for $t \geq T(x_0, \epsilon) + t_0$. The corresponding result holds for the solution $-\bar{\phi}_a(t)$ for all $x_0 \in \mathbb{R}^-$.

Here the family $\mathcal{A} = \{A_a(t), t \in \mathbb{T}\}$ of time dependent sets

$$A_a(t) = \left[-\bar{\phi}_a(t), \bar{\phi}_a(t) \right], \quad t \in \mathbb{R}, \tag{8.8}$$

is the pullback attractor of the Bernoulli equation (8.3). It is also uniformly Lyapunov asymptotically stable and thus a forward atttractor.

8.4 A dissipative ODE

Consider now the scalar nonlinear ODE

$$\frac{dx}{dt} = f(x) + \cos t, \qquad (8.9)$$

where $f : \mathbb{R} \to \mathbb{R}$ is continuously differentiable and satisfies the *dissipativity condition*

$$xf(x) \leq K - L|x|^2, \qquad \forall x \in \mathbb{R},$$

with constants $K \geq 0$ and $L > 0$.

The vector field in the ODE (8.9)

$$\bar{f}(x,t) = f(x) + \cos t$$

also satisfies the dissipativity condition uniformly in $t \in \mathbb{R}$:

$$x\bar{f}(x,t) = xf(x) + x\cos t$$

$$\leq \left(K - L|x|^2 \right) + \left(\frac{L}{2}|x|^2 + \frac{2}{L}|\cos t|^2 \right)$$

$$\leq \left(K + \frac{2}{L} \right) - \frac{L}{2}|x|^2 = \bar{K} - \bar{L}|x|^2$$

with

$$\bar{K} := K + \frac{2}{L}, \qquad \bar{L} := \frac{L}{2}.$$

Thus the solution of the ODE (8.9) with initial value $x(t_0) = x_0$ satisfies

$$\frac{d}{dt}|x(t)|^2 = 2x(t)\bar{f}(x(t),t) \leq 2\bar{K} - 2\bar{L}|x(t)|^2,$$

so

$$|x(t)|^2 \leq |x_0|^2 e^{-2\bar{L}(t-t_0)} + \frac{\bar{K}}{\bar{L}} \left(1 - e^{-2\bar{L}(t-t_0)} \right) \leq 1 + \frac{\bar{K}}{\bar{L}} =: R^2$$

for $t - t_0 \geq T(x_0)$, where

$$T(x_0) := \frac{1}{2\bar{L}} \max \left\{ 0, \ln |x_0|^2 \right\}.$$

For a family $\mathcal{D} = \{D_t : t \in \mathbb{T}\}$ of bounded subsets and the initial values $x_0 \in D_{t_0}$ at time t_0 we take

$$T_{\mathcal{D},t_0} := \sup_{x_0 \in D_{t_0}} T(x_0) < \infty.$$

Then the familiy $\mathcal{B}$ of identical sets $B_t \equiv B_R := [-R, R]$, where R is defined above, is pullback (and also forward) absorbing as well as positive invariant.

Thus the process generated by the ODE (8.9) has a pullback attractor $\mathcal{A}$ with subsets $A_t \subseteq B_R$ for every $t \in \mathbb{R}$.

However, we cannot say anything more about the sets A_t without extra information about the function $f(x)$ in the ODE (8.9).

8.5   One-sided dissipative Lipschitz condition

Let us now assume that the function $f : \mathbb{R} \to \mathbb{R}$ in the ODE (8.9) is continuously differentiable and satisfies the *one-sided dissipativity Lipschitz condition*

$$(x - y)(f(x) - f(y)) \leq -L|x - y|^2, \qquad \forall x, y \in \mathbb{R},$$

with a constant $L > 0$.

An example is the scalar function $f(t, x) = -x - x^3 + 1$ since

$$(x - y)(f(x) - f(y)) = -(x - y)^2 - (x - y)\left(x^3 - y^3\right)$$

$$= -(x - y)^2 - (x - y)^2 \left(x^2 + xy + y^2\right)$$

$$= -(x - y)^2 - (x - y)^2 \left(\left(x + \frac{1}{2}y\right)^2 + \frac{3}{4}y^2\right)$$

$$\leq -(x - y)^2.$$

Then f also satisfies the dissipativity condition above. Taking $y = 0$ gives

$$x(f(x) - f(0)) \leq -L|x|^2,$$

so

$$xf(x) \leq -L|x|^2 + xf(0)$$

$$\leq \frac{2}{L}|f(0)|^2 - \frac{L}{2}|x|^2 = \hat{K} - \hat{L}|x|^2$$

with

$$\hat{K} = \frac{2}{L}|f(0)|^2, \qquad \hat{L} = \frac{L}{2}.$$

Thus by Theorem 7.1 the ODE (8.9) has a pullback attractor $\mathcal{A}$ with component subsets $A_t \subseteq B_R := [-R, R]$, where R is defined by

$$R^2 = 1 + \frac{\bar{K}}{\bar{L}} = 1 + \frac{\hat{K} + \frac{2}{\hat{L}}}{\frac{\hat{L}}{2}} = 1 + \frac{2\hat{K}}{\hat{L}} + \frac{4}{\hat{L}^2}$$

$$= 1 + \frac{2\frac{2}{L}|f(0)|^2}{\frac{L}{2}} + \frac{4}{\left(\frac{L}{2}\right)^2} = 1 + \frac{8}{L^2}\left(|f(0)|^2 + 2\right).$$

Now the difference of two solutions of the ODE (8.9) satisfies

$$\frac{d}{dt}(x(t) - y(t)) = (f(x(t)) - f(y(t))),$$

so

$$\frac{d}{dt}|x(t) - y(t)|^2 = 2(x(t) - y(t))(f(x(t)) - f(y(t))) \leq -2L|x(t) - y(t)|^2$$

which implies that

$$|x(t) - y(t)|^2 \leq |x_0 - y_0|^2 e^{-2L(t - t_0)},$$

and hence that
$$|x(t) - y(t)| \leq |x_0 - y_0| \, e^{-L(t-t_0)}.$$
Thus inside the interval B_R the ODE (8.9) satisfies the *uniform strict contractivity condition*
$$|x(t) - y(t)| \leq 2Re^{-L(t-t_0)}. \tag{8.10}$$
This uniform strict contractivity condition allows to say much more about the pullback attractor $\mathcal{A}$ with component subsets A_t in B_R. If we did not know the result of Theorem 7.2, then we could use the following result.

Proposition 8.1. *When the function $f(x)$ in the ODE (8.9) satisfies the one-sided dissipativity Lipschitz condition, then the pullback attractor consists of a single entire solution and is also forward attracting.*

Proof. Suppose that the pullback attractor does not consist of singleton component sets. Then there are x', $x'' \in A_\tau$ for some $\tau \in \mathbb{R}$ such that $|x' - x''| \geq \varepsilon_0$ for some $\varepsilon_0 > 0$.

Since the pullback attractor $\mathcal{A}$ is ϕ-invariant for the process ϕ generated by the ODE (8.9) for every $t_0 < \tau$ there exist points $x'(t_0)$, $x''(t_0) \in A_{t_0}$ such that
$$\phi(\tau, t_0, x'(t_0)) = x', \qquad \phi(\tau, t_0, x''(t_0)) = x''.$$
By the uniform contractivity condition (8.10) we then obtain
$$\varepsilon_0 \leq |x' - x''| \leq |\phi(\tau, t_0, x'(t_0)) - \phi(\tau, t_0, x''(t_0))|$$
$$\leq 2Re^{-L(\tau-t_0)} \leq \frac{1}{2}\varepsilon_0$$
for
$$t_0 \leq \tau - \frac{1}{2L}\ln 2R.$$
This is a contradiction, so we must have $x' = x''$, i.e., the components A_t are singleton sets, so the pullback attractor consists of a single entire solution.

Let this entire solution be denoted by $e(\cdot)$ so $A_t = \{e(t)\}$ for all $t \in \mathbb{R}$. Then, for any other solution $x(\cdot)$ of the ODE (8.9) the uniform strict contractivity condition (8.10) gives
$$|x(t) - e(t)| \leq |x(t_0) - e(t_0)| \, e^{-L(\tau-t_0)} \to 0 \quad \text{as } t \to \infty.$$
Thus the family of sets $A_t = \{e(t)\}$ is also a forward attractor. $\qquad\square$

Remark 8.1. The one-sided dissipativity Lipschitz condition is a very special case. It means that an autonomous ODE
$$\frac{dx}{dt} = f(x)$$
has a unique globally asymptotically stable steady state solution. A nonlinear example is $f(x) = -x - x^3 + 1$.

The nonautonomous ODE (8.9) obtained by adding $\cos t$ to the right hand side is uniformly strictly contracting and has a nonautonomous attractor consisting of singleton sets, which is both pullback and forward attracting.

Chapter 9

Attractors of skew product flows

Let (P, d_P) and (X, d_X) be complete metric spaces.

Recall that a skew product flow (θ, φ) on $P \times X$ with the time set $\mathbb{T}$ consists of *autonomous dynamical system* $\theta = \{\theta_t\}_{t \in \mathbb{T}}$ on the base space P, which is called the driving system, and a *cocycle mapping* φ on the state space X w.r.t. θ, which describes the dynamics of the state variable.

9.1 Invariant sets and attractors

We saw in Chapter 3 that a process is a special case of a skew product flow with $P = \mathbb{T}$ and θ_t the left shift operator defined by $\theta_t(t_0) = t - t_0$. In addition, we saw that a skew product flow (θ, φ) is an autonomous semi-dynamical system Π on the product space $\mathfrak{X} = P \times X$.

These facts are useful when we consider invariant sets and attractors of skew product flows. They help to motivate the following definitions.

Definition 9.1. A family $\mathcal{A} = \{A_p, p \in P\}$ of nonempty subsets A_p of X is said to be *φ-invariant* for a skew product flow (θ, φ) on $P \times X$ if

$$\varphi(t, p, A_p) = A_{\theta_t(p)} \qquad \text{for all } p \in P \text{ and } t \in \mathbb{T}^+.$$

It is called *φ-positive invariant* if

$$\varphi(t, p, A_p) \subseteq A_{\theta_t(p)} \qquad \text{for all } p \in P \text{ and } t \in \mathbb{T}^+.$$

The essential difference with the definition for a process is that here we use the state of the driving system instead on the initial time.

Similarly to processes we have two types of attractors for skew product flows, pullback and forward attractors. (Later we will consider a third type that comes from the corresponding autonomous semi-dynamical system Π on the state space $\mathfrak{X} = P \times X$).

Definition 9.2. A family $\mathcal{A} = \{A_p, p \in P\}$ of φ-invariant nonempty compact subsets of X is called a *pullback attractor* if it pullback attracts families $\mathcal{D} =$

$\{D_p, p \in P\}$ of nonempty bounded subsets of X, i.e.,

$$\lim_{t \to \infty} \text{dist}_X \left(\varphi(t, \theta_{-t}(p), D_{\theta_{-t}(p)}), A_p \right) = 0 \qquad \text{for each } p \in P.$$

It is called a *forward attractor* if it forward attracts all bounded subsets D of X, i.e.,

$$\lim_{t \to \infty} \text{dist}_X \left(\varphi(t, p, D_p), A_{\theta_t(p)} \right) = 0 \qquad \text{for each } p \in P.$$

Also, as for a process, the existence of a pullback attractor for skew product flow is ensured by that of a pullback absorbing family. As in Remark 7.1 for processes we first restrict to uniformly bounded families of bounded subsets, but this will be generalised later in this chapter to tempered families.

Definition 9.3. A family $\mathcal{B} = \{B_p, p \in P\}$ of nonempty subsets of X is called a *pullback absorbing family* for a skew product flow (θ, φ) on $P \times X$ if for each $t \in P$ and every uniformly bounded family $\mathcal{D} = \{D_p, p \in P\}$ of nonempty bounded subsets of X there exists a $T_{p,\mathcal{D}} \in \mathbb{T}^+$ such that

$$\varphi \left(t, \theta_{-t}(p), D_{\theta_{-t}(p)} \right) \subseteq B_p \qquad \text{for all } t \geq T_{p,\mathcal{D}}.$$

Remark 9.1. Note the difference in the definition of *pullback attraction* here. For a process we start earlier at time τ and finish at the fixed time t, whereas for a skew product flow the driving system starts at the earlier state $\theta_{-t}(p)$ and ends at time t latter at the fixed state $\theta_0(p) = p$. Essentially the driving system in a skew product flow is used instead of the actual value of time, while the time variable in the cocycle mapping just indicates the time since starting at a given state of the driving system.

The proof of the following theorem here is similar to that of Theorem 7.1, at least for a positively invariant absorbing family, so will be omitted. For a proof see [Kloeden and Rasmussen (2011), Theorem 3.20]. As with Theorem 7.1, the assumption that the pullback absorbing family consists of compact sets restricts its usefulness to finite dimensional state spaces such as $\mathbb{R}^d$. An analogous result holds for more general spaces when the absorbing family consists of closed and bounded sets and the cocycle mapping satisfies a pullback asymptotic compactness property.

Theorem 9.1 (Existence of a pullback attractor). *Let (P, d_P) and (X, d_X) be complete metric spaces and suppose that a skew product flow (θ, φ) on $P \times X$ with the time set $\mathbb{T}$ has a pullback absorbing family $\mathcal{B} = \{B_p, p \in P\}$ of nonempty compact sets.*

Then (θ, φ) has a pullback attractor $\mathcal{A} = \{A_p, p \in P\}$ with component subsets determined by

$$A_p = \bigcap_{t \leq 0} \overline{\bigcup_{s \geq t} \varphi \left(t, \theta_{-t}(p), B_{\theta_{-t}(p)} \right)} \qquad \text{for each } p \in P. \tag{9.1}$$

If $\mathcal{B}$ is φ-positively invariant then

$$A_p = \bigcap_{t \leq 0} \varphi\left(t, \theta_{-t}(p), B_{\theta_{-t}(p)}\right) \qquad \text{for each } p \in P.$$

Moreover, $\mathcal{A}$ is unique if the components sets are uniformly bounded.

The possible existence and properties of forward attraction and forward attractors will be discussed later in this and the following chapters. Note for now that if the pullback attractor is uniformly pullback attracting, i.e., if

$$\lim_{t \to \infty} \sup_{p \in P} \text{dist}_X \left(\varphi(t, \theta_{-t}(p), D_{\theta_{-t}(p)}), A_p\right) = 0 \qquad \text{for each } p \in P,$$

then it is uniformly forward attracting, since writing $q = \theta_{-t}(p)$,

$$\sup_{p \in P} \text{dist}_X \left(\varphi(t, \theta_{-t}(p), D_{\theta_{-t}(p)}), A_p\right) = \sup_{q \in P} \text{dist}_X \left(\varphi(t, q, D_q), A_{\theta_t(q)}\right).$$

In this case this uniform pullback/forward attractor is called a *uniform (nonautonomous) attractor*.

9.2 Examples

We consider some examples of pullback attractors of skew product flows, first for continuous time and then for discrete time.

9.2.1 *Continuous-time example*

Consider again the scalar ODE

$$\frac{dx}{dt} = -x + \cos t.$$

As in Example 5.4 define P to be the *hull* of the functions $\cos(\cdot)$, i.e.,

$$P = \bigcup_{0 \leq \tau \leq 2\pi} \cos(\tau + \cdot),$$

which is a compact metric space with the metric induced by the supremum norm

$$\rho(p_1, p_2) = \sup_{t \in \mathbb{R}} |p_1(t) - p_2(t)|.$$

In addition, let $\theta_t : P \to P$ be the left shift operator $\theta_t(\cos(\cdot)) = \cos(t + \cdot)$, so $\theta_{-\tau}(\cos(\cdot)) = \cos(-\tau + \cdot)$. This shift operator is continuous in the above metric.

Now write p_0 for an element of P. The ODE

$$\frac{dx}{dt} = -x + p_0(t)$$

has a unique solution $x(t) = x(t, p_0, x_0)$ with initial value $x(0) = x_0$ given by

$$x(t) = x_0 e^{-t} + e^{-t} \int_0^t e^s p_0(s) \, ds.$$

Then, replacing $p_0(\cdot)$ by $\theta_{-\tau}(p_0(\cdot))$ for some $\tau > 0$ we have

$$x\left(\tau, \theta_{-\tau}(p_0), x_0\right) = x_0 e^{-\tau} + e^{-\tau} \int_0^{\tau} e^s \theta_{-\tau}(p_0(s))\, ds$$

$$= x_0 e^{-\tau} + \int_{-\tau}^0 e^{\rho} p_0(\rho)\, d\rho.$$

with the change of variable $\rho = s - \tau$, so $d\rho = ds$.

We take pullback convergence and obtain the single limit point

$$\lim_{\tau \to \infty} x(t, \theta_{-\tau}(p_0), x_0) = \int_{-\infty}^0 e^{\rho} p_0(\rho)\, d\rho.$$

Remark 9.2. This pullback limit holds for $x_0 \in D$ for any bounded subset D of X because the $x_0 e^{-t}$ term is bounded above by $\|D\| e^{-t}$, where $\|D\| := \sup_{x_0 \in D} \|x_0\| < \infty$. In the next section we will discuss some complications that may arise if we consider arbitrary families $\mathcal{D} = \{D_p, p \in P\}$ nonempty bounded subsets of X.

Now for any two solutions x_1 and x_2 we have

$$\frac{d}{dt}\left(x_1(t) - x_2(t)\right) = -\left(x_1(t) - x_2(t)\right),$$

so

$$\left|(x_1(t) - x_2(t)\right| \leq \left|x_1(0) - x_2(0)\right| e^{-t}.$$

Hence any two solutions converge forward in time together. This means that the pullback attractor consists of singleton subsets

$$A_{p_0} = \left\{ \int_{-\infty}^0 e^{\rho}\, p_0(\rho)\, d\rho \right\}$$

that are also forwards attracting. As we saw earlier, this system is uniformly strictly contracting.

An alternative driving system

Define the function $p_0 : \mathbb{R} \to \mathbb{R}$ by $p_0(t) = \cos(p_0 + t)$ for $t \in \mathbb{R}$ with $p_0 \in [0, 2\pi]$, i.e., with the second p_0 denoting the phase. Then we have

$$\int_{-\infty}^0 e^{\rho} p_0(\rho)\, d\rho = \int_{-\infty}^0 e^{\rho} \cos(p_0 + \rho)\, d\rho$$

$$= e^{-p_0} \int_{-\infty}^0 e^{p_0 + \rho} \cos(p_0 + \rho)\, d\rho$$

$$= e^{-p_0} \int_{-\infty}^{p_0} e^{r} \cos(r)\, dr = \frac{1}{2}\left(\cos p_0 + \sin p_0\right).$$

This means that

$$A_{p_0} = \left\{ \frac{1}{2} \left(\cos p_0 + \sin p_0 \right) \right\}, \qquad p_0 \in [0, 2\pi].$$

Thus it seems that we can represent the driving system θ_t on P by $\tilde{\theta}_t(p_0) := p_0 + t \mod 2\pi$ on $\tilde{P} = [0, 2\pi]$. However, $\tilde{\theta}_t$ is not continuous on $[0, 2\pi]$. To ensure its continuity, we need to take $\tilde{P} = \mathbb{R}/2\pi\mathbb{Z}$, which identifies periodically related points.

9.2.2 *Discrete-time example*

Consider the scalar nonautonomous difference equation

$$x_{n+1} = ax_n + b_n,$$

where $a \in (0, 1)$ and $b_n \in [-B, B]$, i.e., the b_n are uniformly bounded with $|b_n| \le B$.

Define abi-infinite sequence

$$\mathbf{b} = (\ldots, b_{-2}, b_{-1}, b_0, b_1, b_2, \ldots)$$

in $P := [-B, B]^{\mathbb{Z}}$ with the metric

$$\rho(\mathbf{b}, \mathbf{b}') := \sum_{n \in \mathbb{Z}} 2^{-|n|} |b_n - b'_n|.$$

The space (P, ρ) is a compact metric space.

Let $\theta : P \to P$ be the left shift operator defined by

$$\mathbf{b}' := \theta(\mathbf{b}) \qquad \Leftrightarrow \qquad b'_n = b_{n+1} \qquad \text{for all} \quad n \in \mathbb{Z}.$$

Then θ is a homeomorphism, i.e., both it and its inverse are continuous, and $\{\theta^n, n \in \mathbb{Z}\}$ is group on P under composition, so $\theta_n \equiv \theta^n$.

For a given sequence $\mathbf{b}$ the solution mapping is

$$\varphi(n, \mathbf{b}, x_0) = x_n = a^n x_0 + \left(a^{n-1} b_0 + \cdots + a b_{n-2} + b_{n-1} \right)$$

$$= a^n x_0 + \sum_{j=0}^{n-1} a^{n-1-j} b_j \qquad n \ge 1.$$

Hence,

$$\varphi(n, \theta_{-n}(\mathbf{b}), x_0) = a^n x_0 + \sum_{j=0}^{n-1} a^{n-1-j} b_{j-n} \qquad n \ge 1,$$

since $(\theta_{-n}(\mathbf{b}))_j = b_{j-n}$. Relabelling the summation indices, i.e., $k = j - n$, we have

$$\varphi(n, \theta_{-n}(\mathbf{b}), x_0) = a^n x_0 + \sum_{k=-n+1}^{0} a^{-1-k} b_k \qquad n \ge 1.$$

Pullback convergence then gives

$$\lim_{n\to\infty} \varphi\left(n, \theta_{-n}(\mathbf{b}), x_0\right) = \sum_{k=-\infty}^{0} a^{-1-k} b_k =: \alpha(\mathbf{b}).$$

The infinite series convergence here because

$$\left| a^{-1-k} b_k \right| \leq a^{-1+|k|} |b_k| \leq a^{-1+|k|} B \to 0 \quad \text{as } k \to -\infty.$$

Since the difference of any two solutions satisfies

$$x_{n+1} - y_{n+1} = a\left(x_n - y_n\right) \qquad \Leftrightarrow \qquad \left(x_n - y_n\right) = a^n\left(x_0 - y_0\right),$$

they converge together in the forward sense. This means that the pullback attractor consists of singleton component sets

$$A_{\mathbf{b}} = \{\alpha(\mathbf{b})\} = \left\{ \sum_{k=-\infty}^{0} a^{-1-k} b_k \right\}$$

is also a forward attractor. The system here is also uniformly strictly contracting.

Remark 9.3. These examples show that it is usually more complicated to determine the pullback limit for a system formulated as a skew product flow than as a process.

9.3   Basin of pullback attraction system

In Definition 9.3 it was assumed that the pullback and forward attractors attract *all* families $\mathcal{D} = \{D_p, p \in P\}$ nonempty bounded subsets of X. This allows one to take into account nonuniformities which are typical in nonautonomous behaviour.

However, we cannot always take arbitrary bounded subsets D_p of X. As we saw in Remark 9.2, in Example 9.2.1 the term $\|D_{\theta_{-t}(p_0)}\| e^{-t}$ must vanish as $t \to \infty$, which means that the bounds $\|D_{\theta_{-t}(p_0)}\|$ should not grow too quickly. We can overcome this by restricting to families for which these bounds grow at most subexponentially.

Definition 9.4. A family $\mathcal{D} = \{D_p, \ p \in P\}$ of nonempty bounded subsets D_p of X is said to have *subexponential growth* if

$$\limsup_{|t|\to\infty} \|D_{\theta_{-t}(p_0)}\| e^{\epsilon|t|} = 0 \qquad \forall \epsilon > 0.$$

In this case it is called a *tempered* family.

This may also be too general. Often we need to restrict to particular classes of such families called *universes*.

Definition 9.5. A *universe* $\mathfrak{D}_{att}$ of a skew product flow (θ, φ) on $P \times X$ is collection of families $\mathcal{D} = \{D_p, \ p \in P\}$ of nonempty bounded subsets D_p of X which is closed under inclusion, i.e., $\mathcal{D}^{(1)} = \left\{ D_p^{(1)}, p \in P \right\} \in \mathfrak{D}_{att}$ if $\mathcal{D}^{(2)} = \left\{ D_p^{(2)}, \ p \in P \right\} \in \mathfrak{D}_{att}$ and $D_p^{(1)} \subseteq D_p^{(2)}$ for all $p \in P$.

The definition of a pullback attractor then needs to be modified.

Definition 9.6. A φ-invariant family $\mathcal{A} = \{A_p, \ p \in P\}$ of nonempty compact subsets of X is called a pullback attractor with respect to a universe $\mathfrak{D}_{att}$ of a skew product flow (θ, φ) if

$$\lim_{t \to \infty} \text{dist}\left(\varphi(t, \theta_{-t}(p), D_{\theta_{-t}(p)}), A_p\right) = 0 \tag{9.2}$$

for all $p \in P$ and all $\mathcal{D} = \{D_p, \ p \in P\} \in \mathfrak{D}_{att}$.

Such a universe $\mathfrak{D}_{att}$ is often called the *basin of attraction* system of the pullback attractor $\mathcal{A}$.

Obviously, $\mathcal{A} \in \mathfrak{D}_{att}$. In fact, $A_p \subset \text{int}\,\mathfrak{D}_{att}(p)$ for each $p \in P$, where

$$\mathfrak{D}_{att}(p) := \bigcup_{\mathcal{D} = \{D_p, \ p \in P\} \in \mathfrak{D}_{att}} D_p \subset X$$

for each $p \in P$.

Universes are also useful for discussing *local* pullback or forward attractors. As a simple example, consider the autonomous semi-dynamical system on $X = \mathbb{R}^1$ generated by the scalar ODE

$$\frac{x}{dt} = x(1 - x). \tag{9.3}$$

This can be formulated as a skew product flow on a base space P consisting of just one point.

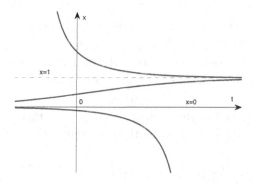

Fig. 9.1   Typical solutions of the ODE (9.3).

The compact invariant set $[0, 1]$ is not a global attractor, but it is a local attractor in the nonnegative subspace $\mathbb{R}^1_+ = \{x \in \mathbb{R}^1; x \geq 0\}$. The basin of attraction system here consists of all families consisting of a single bounded subset of $\mathbb{R}^1_+$.

9.4   Upper semi-continuity of the set-valued mapping $p \mapsto A_p$

It follows [Aubin and Franksowka (1990)] from the compactness of the sets A_p and the continuity of the single-valued mapping $(t, p, x) \mapsto \varphi(t, p, x)$ that the set-valued mapping $t \mapsto \varphi(t, p, A_p)$ is continuous for each $p \in P$. Hence, by the φ-invariance of the pullback attractor the set-valued mapping

$$t \mapsto A_{\theta_t(p)} = \varphi(t, p, A_p)$$

is continuous for each $p \in P$.

Theorem 9.2. *Let $\mathcal{A} = \{A_p, p \in P\}$ be a pullback attractor of a skew product flow (θ, φ) on $P \times X$ with the time set $\mathbb{T}$ and suppose that*

$$A(P) := \overline{\bigcup_{p \in P} A_p}$$

is compact.

    Then the set-valued mapping $p \mapsto A_p$ is upper semi-continuous in the sense that

$$\mathrm{dist}_X(A_q, A_p) \to 0 \quad \text{as } q \to p.$$

    On the other hand, if P is compact and the set-valued mapping $p \mapsto A_p$ is upper semi-continuous, then $A(P)$ is compact.

Proof. Since $A(P)$ is compact, the pullback attractor component sets are uniformly bounded, so the pullback attractor is unique.

    Suppose that the set-valued mapping $p \mapsto A_p$ is *not* upper semi-continuous. Then there exists $\varepsilon_0 > 0$ and a sequence $p_n \to p_0$ in P such that

$$\mathrm{dist}_X(A_{p_n}, A_{p_0}) \geq 3\varepsilon_0 \quad \text{for all } n \in \mathbb{N}.$$

Since the sets A_{p_n} are compact, there exists points $a_{p_n} \in A_{p_n}$ for each $n \in \mathbb{N}$ such that

$$\mathrm{dist}_X(a_{p_n}, A_{p_0}) = \mathrm{dist}_X(A_{p_n}, A_{p_0}) \geq 3\varepsilon_0 \quad \text{for all } n \in \mathbb{N}. \tag{9.4}$$

    By pullback attraction, for every bounded subset B of X we have

$$\mathrm{dist}_X(\varphi(t, \theta_{-t}(p_0), B), A_{p_0}) \leq \varepsilon_0 \quad \text{for all } t \geq T_{B,p_0,\varepsilon_0}.$$

We will use this for the bounded set $B = A(P)$. By φ-invariance of the pullback attractors, there exists $b_n \in A_{\theta_{-t}(p_n)}$ (with a fixed t here) for each $n \in \mathbb{N}$ such that

$$\varphi(t, \theta_{-t}(p_0), b_n) = a_n.$$

Now the b_n are contained in the compact set $A(P)$, so there is a convergent subsequence $b_{n'} \to \bar{b} \in A(P)$. Finally, by the continuity of $p \mapsto \theta_{-t}(p)$ and $(p, x) \mapsto \varphi(t, p, x)$ we have

$$d_X(\varphi(t, \theta_{-t}(p_{n'}), b_{n'}), \varphi(t, \theta_{-t}(p_0), \bar{b})) \leq \varepsilon_0 \quad \text{for all large enough } n'.$$

Thus

$$\mathrm{dist}_X(A_{p_{n'}}, A_{p_0}) = \mathrm{dist}_X(\varphi(t, \theta_{-t}(p_{n'}), b_{n'}), A_{p_0})$$

$$\leq d_X(\varphi(t, \theta_{-t}(p_{n'}), b_{n'}), \varphi(t, \theta_{-t}(p_0), \bar{b}))$$

$$+ \mathrm{dist}_X(\varphi(t, \theta_{-t}(p_0), \bar{b}), A_{p_0}) \leq 2\varepsilon_0,$$

which contradicts the inequality (9.4). Thus $p \mapsto A_p$ is upper semi-continuous.

    The remaining assertion of the theorem holds because the image of a compact subset under an upper semi-continuous set-valued mapping with compact values is compact. (See [Aubin and Franksowka (1990)] for properties of set-valued mappings.) $\square$

9.4.1 *The set-valued mapping $p \mapsto A_p$ need not be continuous*

We have just shown that, under certain conditions, the set-valued mapping $p \mapsto A_p$ is upper semi-continuous. Here we show that it need not be continuous.

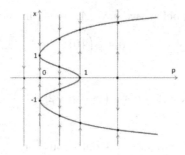

Fig. 9.2 Supercritical bifurcation in the autonomous ODE (9.5) with parameter p.

Consider that autonomous scalar ODE

$$\frac{dx}{dt} = -x \left(x^4 - 2x^2 + 1 - p \right) \qquad (9.5)$$

with a parameter $p \in P = [-2, 2]$.

The steady state $\bar{x}_p = 0$, which exists for all $p \in P$, under goes a *subcritical bifurcation* at $p = 0$. It has the steady state solutions

- $\bar{x}_p = 0$ for $p < 0$,

- $\bar{x}_p = 0$, $\pm\sqrt{1 + \sqrt{p}}$ and $\pm\sqrt{1 - \sqrt{p}}$ for $0 < p < 1$,

- $\bar{x}_p = 0$ and $\pm\sqrt{1 + \sqrt{p}}$ for $1 \leq p$.

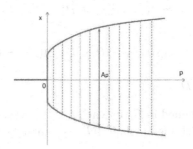

Fig. 9.3 Global attractor A_p in the autonomous ODE (9.5) with parameter p.

The global attractors (autonomous) are

$$A_p = \begin{cases} \{0\} & \text{for } p < 0, \\ \\ \left[-\sqrt{1 + \sqrt{p}}, \sqrt{1 + \sqrt{p}} \right] & \text{for } p \geq 0. \end{cases}$$

Now consider the *trivial* nonautonomous dynamical system with the constant driving system $\theta_t(p) \equiv p$ for each $p \in P = [-2, 2]$ and all $t \in \mathbb{R}$. The pullback attractor $\mathcal{A}$ has these A_p as its component sets: since $\theta_t(p) \equiv p$, the dynamics just stays on the same A_p for all $t \in \mathbb{R}$.

For $p \to 0^-$, i.e., from below, we have $A_p \to A_0$ in the Hausdorff semi-distance

$$\text{dist}_{\mathbb{R}}\left(A_p, A_0\right) = \text{dist}_{\mathbb{R}}\left(\{0\}, [-\sqrt{2}, \sqrt{2}]\right) = 0,$$

whereas

$$\text{dist}_{\mathbb{R}}\left(A_0, A_p\right) = \text{dist}_{\mathbb{R}}\left([-\sqrt{2}, \sqrt{2}], \{0\}\right) = \sqrt{2} \neq 0$$

for all $p < 0$.

The convergence is *only* upper semi-continuous convergence and not lower semi-continuous convergence and hence not continuous convergence.

9.5   A weak form of forward convergence

Pullback attractors are, in general, <u>not</u> forward attractors. We saw this for processes and it is also true for skew product flows. But if the base space P is compact, then we have the following partial result on forward convergence when we have a pullback attractor. Essentially, the set $A(P)$ contains all of the limiting dynamics in the forwards sense, but it is not an attractor as it is not invariant. (More will be said on this topic in Chapter 11.)

Theorem 9.3. *Assume, in addition to the asumptions of Theorem 9.1, that the base space P is compact and that the pullback absorbing family $\mathcal{B} = \{B_p, p \in P\}$ is uniformly bounded by a compact set, i.e., there exists a compact set C of X with $B_p \subseteq C$ for each $p \in P$.*

Then the pullback attractor $\mathcal{A} = \{A_p, p \in P\}$ of the skew product flow (θ, φ) on $P \times X$ satisfies

$$\lim_{t \to \infty} \sup_{p \in P} \text{dist}_X\left(\varphi\left(t, p, D\right), A(P)\right) = 0 \qquad (9.6)$$

for every bounded subset D of X.

Proof.   Since A_p is contained in B_p for each $p \in P$ so we have $A_p \subseteq C$ for each $p \in P$. Hence $A(P) \subseteq C$, which means that $A(P)$ is compact, so the pullback attractor is unique.

Suppose that the limit (9.6) does not hold. Then there exist a $\varepsilon_0 > 0$ and sequences $t_j \to \infty$ in $\mathbb{T}$, $\hat{p}_j \in P$ and $x_j \in C$ such that

$$\text{dist}_X\left(\varphi\left(t_j, \hat{p}_j, x_j\right), A(P)\right) \geq 3\varepsilon_0. \qquad (9.7)$$

Set $p_j = \theta_{t_j}(\hat{p}_j)$. By the compactness of P there exists a convergent subsequence $\hat{p}_{j'} \to p_0 \in P$.

From the pullback attraction for $\tau \geq 0$ large enough

$$\text{dist}_X \left(\varphi \left(\tau, \theta_{-\tau}(p_0), C \right), A_{p_0} \right) \leq \varepsilon_0.$$

The cocycle property gives

$$\varphi \left(t_j, p_j, x_j \right) = \varphi \left(\tau, \theta_{-\tau}(p_j), \varphi \left(t_j - \tau, \theta_{-t_j}(p_j), x_j \right) \right)$$

for $t_j > \tau$.

By the pullback absorbing property of $\mathcal{B}$ it follows that

$$\varphi \left(t_j - \tau, \theta_{-t_j}(p_j), x_j \right) \subseteq B_{\theta_{-\tau}(p_j)} \subseteq C.$$

Since C is compact there exists a convergent (sub)sequence

$$z_{j''} = \varphi \left(t_{j''} - \tau, \theta_{-t_{j''}}(p_{j''}), x_{j''} \right) \to z_0 \in C.$$

The continuity of $p \mapsto \theta_t p$ and $(p, x) \mapsto \varphi(t, p, x)$ then implies that

$$\text{dist}_X \left(\varphi \left(\tau, \theta_{-\tau}(p_{j''}), z_{j''} \right), \varphi \left(\tau, \theta_{-\tau}(p_0), z_0 \right) \right) \leq \varepsilon_0$$

for $j'' \geq N(\varepsilon_0)$, i.e., large enough. Hence

$$\text{dist}_X \left(\varphi \left(t_{j''}, \theta_{-t_{j''}}(p_{j''}), x_{j''} \right), A_{p_0} \right)$$

$$\leq \text{dist}_X \left(\varphi \left(t_{j''}, \theta_{-t_{j''}}(p_{j''}), x_{j''} \right), \varphi \left(\tau, \theta_{-\tau}(p_0), z_0 \right) \right) + \text{dist}_X \left(\varphi \left(\tau, \theta_{-\tau}(p_0), z_0 \right), A_{p_0} \right)$$

$$= \text{dist}_X \left(\varphi \left(\tau, \theta_{-\tau}(p_{j''}), z_{j''} \right), \varphi \left(\tau, \theta_{-\tau}(p_0), z_0 \right) \right) + \text{dist}_X \left(\varphi \left(\tau, \theta_{-\tau}(p_0), z_0 \right), A_{p_0} \right)$$

$$\leq 2\varepsilon_0.$$

Therefore

$$2\varepsilon_0 > \text{dist}_X \left(\varphi \left(t_{j''}, \theta_{-t_{j''}}(p_{j''}), x_{j''} \right), A_{p_0} \right) = \text{dist}_X \left(\varphi \left(t_{j''}, \hat{p}_{j''}, x_{j''} \right), A_{p_0} \right)$$

$$\geq \text{dist}_X \left(\varphi \left(t_{j''}, \hat{p}_{j''}, x_{j''} \right), A(P) \right) > 3\varepsilon_0,$$

which is a contradiction, so the asserted convergence must hold. $\square$

Example 9.1. Suppose that $\theta_t : P \to P$ is T-periodic, i.e., there is a $T > 0$ such that $\theta_T(p) = p$ for all $p \in P$, and that P consists of single θ trajectory, i.e., there is a $p_0 \in P$ such that $P = \{\theta_t(p_0), 0 \leq t \leq T\}$. Hence P is compact.

In addition, suppose that the pullback attractor has singleton component sets $A_p = \{a_p\}$, so $t \mapsto a_{\theta_t}$ is T-periodic. Then $A(P) = \{a_{\theta_t(p_0)}, 0 \leq t \leq T\}$ represents the geometric curve in X of the periodic solution forming the pullback attractor. Theorem 9.3 says that other solutions converge forwards in time to this geometric limit cycle.

However, the family $\{A_p, p \in P\}$ need not be a forward attractor since the convergence to the point at the same time instant on the curve, i.e., $d_X(\phi(t, p, x_0), a_{\theta_t(p)}) \to 0$, need not hold $t \to \infty$. (It will hold if, for example, the system is uniformly strictly attracting as in the examples in Section 9.2).

9.6   Another type of attractor for skew product flows

We consider some more results on the relationship between the attractors of a skew product flow and its representation as an autonomous semi-dynamical system.

Let (θ, φ) be a skew product flow on $P \times X$ with a time set $\mathbb{T}$ and let Π be the corresponding autonomous semi-dynamical system on $\mathfrak{X} = P \times X$, i.e., defined by

$$\Pi\left(t, (p_0, x_0)\right) = \left(\theta_t(p_0), \varphi(t, p_0, x_0)\right).$$

We use the metric $d_{\mathfrak{X}}$ on $\mathfrak{X}$ defined by

$$d_{\mathfrak{X}}\left((p_1, x_1), (p_2, x_2)\right) = d_P\left(p_1, p_2\right) + d_X\left(x_1, x_2\right).$$

Proposition 9.1. (Uniform convergence of global attractors)   *Suppose that $\mathcal{A} = \{A_p, p \in P\}$ is a uniform attractor for the skew product flow (φ, θ), i.e., uniformly attracting in both forward and pullback senses. Also, assume that P and the set*

$$A(P) = \overline{\bigcup_{p \in P} A_p}$$

are compact in X.

   Then the union

$$\mathfrak{A} = \bigcup_{p \in P} \{p\} \times A_p$$

is the global attractor of the autonomous semi-dynamical system Π.

Proof   The Π-invariance of $\mathfrak{A}$ follows by the φ-invariance of $\mathcal{A}$ and the θ-invariance of P via

$$\Pi\left(t, \mathfrak{A}\right) = \bigcup_{p \in P} \{\theta_t(p)\} \times \varphi\left(t, p, A_p\right)$$

$$= \bigcup_{p \in P} \{\theta_t(p)\} \times A_{\theta_t(p)} = \bigcup_{q \in P} \{q\} \times A_q = \mathfrak{A}$$

since by the θ-invariance of P for each $\theta_t(p)$ there is a $q \in P$ with $q = \theta_t(p)$.

   Since $\mathcal{A}$ is a pullback attractor and $A(P)$ is compact, the set-valued mapping $p \mapsto A_p$ is upper semi-continuous. This means that the mapping

$$p \mapsto F(p) := \{p\} \times A_p$$

is also upper semi-continuous. Hence $F(P) = \mathfrak{A}$ is a compact subset of $\mathfrak{X} = P \times X$. See Aubin and Frankowska [Aubin and Franksowka (1990)] for the properties of set-valued mappings.

   Moreover, by the definition of the metric $d_{\mathfrak{X}}$ on $\mathfrak{X}$, we have

$$\text{dist}_{\mathfrak{X}}\left(\Pi\left(t, (p, x)\right), \mathfrak{A}\right) = \text{dist}_{\mathfrak{X}}\left((\theta_t(p), \varphi(t, p, x)), \mathfrak{A}\right)$$

$$\leq \text{dist}_{\mathfrak{X}}\left((\theta_t(p), \varphi(t, p, x)), \{\theta_t(p)\} \times A_{\theta_t(p)}\right)$$

$$= \underbrace{\text{dist}_P\left(\theta_t(p), \theta_t(p)\right)}_{=0} + \text{dist}_X\left(\varphi(t, p, x), A_{\theta_t(p)}\right)$$

$$= \text{dist}_X\left(\varphi(t, p, x), A_{\theta_t(p)}\right) \to 0 \quad \text{as } t \to \infty$$

since $\mathcal{A}$ is also a forward attractor. $\square$

Without the assumption forward attraction, the pullback attractor need not give a global attractor of the autonomous semi-dynamical system Π.

Proposition 9.2. *Suppose that $A = \{A_p, p \in P\}$ is a pullback attractor of the skew product flow (θ, φ) on $P \times X$ and that P and $A(P) = \overline{\cup_{p \in P} A_p}$ are compact in X.*
Then $\mathfrak{A} = \cup_{p \in P} \{p\} \times A_p$ is the maximal invariant compact subset in $\mathfrak{X}$ of the autonomous semi-dynamical system Π.

Proof. The compactness and Π-invariance of $\mathfrak{A}$ follow in the same way as in the previous proof.

To prove that the compact invariant set $\mathfrak{A}$ is maximal. Let $\mathfrak{C} = \cup_{p \in P} \{p\} \times C_p$ be another compact Π-invariant subset in $\mathfrak{X} = P \times X$. Then $C = \{C_p, p \in P\}$ is a φ-invariant family of compact subsets of X. By pullback attraction

$$\text{dist}_X (C_p, A_p) = \text{dist}_X \left(\varphi(t, \theta_{-t}(p), C_{\theta_{-t}(p)}), A_p \right)$$

$$\leq \text{dist}_X \left(\varphi(t, \theta_{-t}(p), C(P)), A_p \right) \to 0 \quad \text{as } t \to \infty,$$

where

$$C(P) = \overline{\bigcup_{p \in P} C_p}$$

is a compact subset of X.

Thus $C_p \subseteq A_p$ and

$$\mathfrak{C} = \bigcup_{p \in P} \{p\} \times C_p \subseteq \bigcup_{p \in P} \{p\} \times A_p = \mathfrak{A},$$

i.e., $\mathfrak{C}$ is a subset of $\mathfrak{A}$. Hence $\mathfrak{A}$ is maximal. $\square$

Remark 9.4. The set $\mathfrak{A}$ in the above proposition need not be a global attractor of the autonomous semi-dynamical system Π on $\mathfrak{X} = P \times X$. For an example in terms of processes see Section 10.3, while Cheban, Kloeden and Schmalfuß [Cheban, Kloeden and Schmalfuß (2001)] give an example in terms of skew product flows.

In the other direction we have

Proposition 9.3. *If P is compact and the autonomous semi-dynamical system Π in $\mathfrak{X}$ has a global attractor $\mathfrak{A} = \cup_{p \in P} \{p\} \times A_p$, then the family $A = \{A_p, p \in P\}$ is a pullback attractor for the skew product flow (θ, φ).*

Proof.   The sets P and $K := \overline{\cup_{p \in P} A_p}$ are compact by the compactness of $\mathfrak{A}$. Moreover, $\mathfrak{A} \subseteq P \times K$, which is a compact set. Now

$$\mathrm{dist}_X\left(\varphi(t,p,x), K\right) = \underbrace{\mathrm{dist}_P\left(\theta_t(p), P\right)}_{=0} + \mathrm{dist}_X\left(\varphi(t,p,x), K\right)$$

$$= \mathrm{dist}_{\mathfrak{X}}\left((\theta_t(p), \varphi(t,p,x)), P \times K\right)$$

$$= \mathrm{dist}_{\mathfrak{X}}\left(\Pi\left(t,(p,x)\right), P \times K\right)$$

$$\leq \mathrm{dist}_{\mathfrak{X}}\left(\Pi\left(t, P \times D\right), P \times K\right), \qquad x \in D,$$

$$\leq \mathrm{dist}_{\mathfrak{X}}\left(\Pi\left(t, P \times D\right), \mathfrak{A}\right) \to 0 \quad \text{as } t \to \infty,$$

for any bounded subset D of X (hence $P \times D$ is bounded in $\mathfrak{X}$), since $\mathfrak{A}$ is the global attractor of Π.

Replacing p by $\theta_{-t}(p)$ here we have

$$\lim_{t \to \infty} \mathrm{dist}_X\left(\varphi(t, \theta_{-t}(p), D), K\right) = 0.$$

This says that the skew product flow (θ, φ) is *asymptotically compact*, which is a *sufficient condition* for the existence of a pullback attractor $\mathcal{A}' = \{A'_p, p \in P\}$ with $A'_p \subseteq K$.

From Proposition 9.2 above, $\mathfrak{A}' = \cup_{p \in P}\{p\} \times A'_p$ is a maximal invariant compact subset of $\mathfrak{X}$, but so is the global attractor. This means that $\mathfrak{A}' = \mathfrak{A}$ and hence that $\mathcal{A}' = \mathcal{A}$.   □

PART 3

Forward attractors and attracting sets

Chapter 10

Limitations of pullback attractors of processes

Pullback attractors have been extensively investigated in past decades. Essentially, they describe the behaviour of a system "from the past" and, in general, have little to say about the future behaviour of the system. This will be illustrated here with some simple examples involving *difference equations*, i.e., discrete-time systems.

We will also be seen that the skew product flow formulation often provides more detail about the behaviour of a nonautonomous dynamical system than does the process formulation due to its built-in driving system.

The future behaviour of the system is described by forward attractors or, more generally, forward omega limit sets. These have only become better understood in recent years and will be discussed in the last chapters of the book.

Both pullback and forward attractors or omega limit sets, when they exist, are needed to provide a full description of the dynamical behaviour of a nonautonomous system.

In the simple examples here the pullback attractors are with respect to the pullback attraction of bounded sets or families of uniformly bounded sets.

10.1 An autonomous difference equation

We start with an *autonomous* scalar difference equation

$$x_{n+1} = \frac{\lambda x_n}{1 + |x_n|},$$ (10.1)

where λ is a real-valued parameter with $\lambda \geq 0$. Clearly, $x^* = 0$ is always a *steady state* solution for all $\lambda \geq 0$.

Other steady state solutions are obtained by solving

$$x^* = \frac{\lambda x^*}{1 + |x^*|}$$

for $x^* \neq 0$, i.e., solving $1 + |x^*| = \lambda$ or

$$|x^*| = \lambda - 1.$$

Such solutions exist only if $\lambda \geq 1$. In fact, the point $\lambda^* = 1$ is a *bifurcation point*, where two new steady states

$$x_\lambda^* = \pm (\lambda - 1)$$

arise for $\lambda > \lambda^* = 1$. There are thus two cases:

($\lambda \leq 1$): $x^* = 0$ is the only steady state solution and it is globally asymptotically stable. The global (autonomous) attractor is $A = \{0\}$.

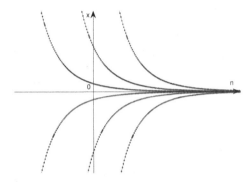

Fig. 10.1   Typical solutions (interpolated) of the difference equation (10.1) with $\lambda \leq 1$.

($\lambda > 1$): $x^* = 0$ is the unstable steady state solution. The other steady states x_λ^* $= \pm (\lambda - 1)$ are locally asymptotically stable with basins of attractors given by $\pm x_0$ > 0. The global attractor, i.e., which attracts all x_0 in $\mathbb{R}$, in this case is

$$A_\lambda = \big[- |x_\lambda^*|, |x_\lambda^*| \big] = [1 - \lambda, \lambda - 1].$$

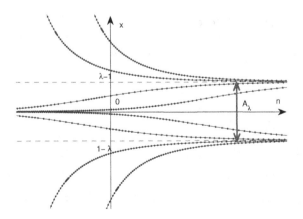

Fig. 10.2   Typical solutions (interpolated) of the difference equation (10.1) with $\lambda > 1$.

10.2 A piecewise autonomous difference equation

Consider the same difference equation (10.1) with the parameter λ switching its value, specifically

$$x_{n+1} = \frac{\lambda_n x_n}{1 + |x_n|}, \qquad \text{where} \quad \lambda_n = \begin{cases} \lambda & \text{for } n \geq 0 \\ \lambda^{-1} & \text{for } n \leq 0 \end{cases} \qquad (10.2)$$

for some $\lambda \geq 1$.

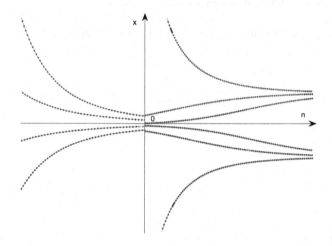

Fig. 10.3 Typical solutions (interpolated) of the switching system (10.2).

This is a nonautonomous difference equation which results from *switching* between the two autonomomous difference equations at $n = 0$.

The only *bounded entire solution* is the zero solution

$$\bar{x}_n \equiv 0 \qquad \text{for all} \quad n \in \mathbb{Z}$$

and the *pullback attractor* consists of the identical singleton sets

$$A_n \equiv \{0\} \qquad \text{for all} \quad n \in \mathbb{Z}.$$

Remark 10.1. The zero solution appears to be *stable* for $n \leq 0$ (it is indeed asymptotically stable for the autonomous difference equation acting for those times), but it is then *unstable* for $n > 0$. For $n \geq 0$, the set $\left[-|x_\lambda^*|, |x_\lambda^*| \right]$ looks like a *global attractor* (it is indeed the global attractor for the autonomous difference equation acting for these times), but it cannot be an attractor for this piecewise system because it is <u>not</u> invariant for $n < 0$.

Obviously the set of pullback attractor points $\bigcup_{n \in \mathbb{Z}} A_n = \{0\}$ is contained in $\left[-|x_\lambda^*|, |x_\lambda^*| \right]$, but the pullback attractor gives little information about the *future* asymptotic dynamics after the switch at $n^* = 0$. In particular, this example has no uniformly bounded forward attractor.[1]

Remark 10.2. One often talks about *asymptotically autonomous* systems in one or both directions as $n \to \pm\infty$ and analyses the "limiting" asymptotic dynamics of the limiting autonomous systems separately.

10.3 Fully nonautonomous system

Let us now consider the nonautonomous difference equation

$$x_{n+1} = \frac{\lambda_n x_n}{1 + |x_n|}, \tag{10.3}$$

where $\{\lambda_n\}_{n\in\mathbb{Z}}$ is the *monotonically increasing* sequence given by

$$\lambda_n = 1 + \frac{0.9n}{1 + |n|}, \qquad \text{for all} \quad n \in \mathbb{Z}.$$

Obviously, $\lambda_n < 1$ for $n < 0$ and $\lambda_n > 1$ for $n > 0$. Note that

$$\lambda_n = 1 - \frac{0.9}{\frac{1}{|n|} + 1} \to 0.1 \quad \text{as} \quad n \to -\infty,$$

for $n < 0$, while

$$\lambda_n = 1 + \frac{0.9}{\frac{1}{n} + 1} \to 1.9 =: \bar{\lambda} \quad \text{as} \quad n \to +\infty$$

for $n > 0$. The convergence here is monotone. Moreover, λ never reaches the limit $\bar{\lambda}$.

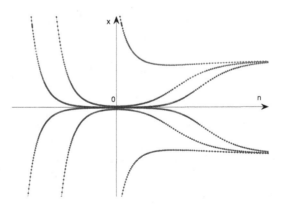

Fig. 10.4   Typical solutions (interpolated) of the system (10.3) with monotonically increasing λ_n.

---

[1]If x_n is any unbounded entire positive solution converging to $x_{\bar{\lambda}}^*$ as $n \to \infty$ then the sets $A_n = [-x_n x_n]$ are forward attracting, but grow unboundedly as $n \to -\infty$.

As above, the zero solution is the only bounded entire solution. It appears to be "asymptotically stable" for $n < 0$ and to become unstable for $n > 0$. The pullback attractor consists of the identical singleton sets

$$A_n \equiv \{0\} \qquad \text{for all} \quad n \in \mathbb{Z}.$$

However, in this case the forward limit points $\pm (\bar{\lambda} - 1)$ are not solutions of the difference equation at all. In particular, they are not entire solutions, so they cannot belong to an attractor, either autonomous or nonautonomous. This example has no uniformly bounded forward attractor (but see the previous footnote).

Remark 10.3. Note that we have used the process formulation of the nonautonomous system in the above analysis without actually explicitly saying so. In this and the previous example the pullback attractor provides only partial information about the full dynamics, in particular, about the future dynamics, of the difference equations.

10.4 More information through skew products

Consider the above nonautonomous difference equation (10.3)

$$x_{n+1} = \frac{\lambda_n x_n}{1 + |x_n|},$$

where $\lambda = \{\lambda_n\}_{n \in \mathbb{Z}}$ is a given strictly increasing sequence with

$$0 < \lambda^- < \cdots < \lambda_n < \lambda_{n+1} < \cdots < \lambda^+$$

where

$$0 < \lambda^- := \lim_{n \to -\infty} \lambda_n < 1 < \lambda^+ := \lim_{n \to +\infty} \lambda_n,$$

that is, with

$$0 < \lambda^- < 1 < \lambda^+.$$

Let Λ be the space of real valued bi-infinite sequences $\rho = \{\rho_n\}_{n \in \mathbb{Z}}$ consisting of the constant sequences $\rho^\pm$ with components $\rho_n^- \equiv \lambda^-$ and $\rho_n^+ \equiv \lambda^+$ as well as all shifts $\theta^k \lambda$, $k \in \mathbb{Z}$, of the given sequence $\lambda = \{\lambda_n\}_{n \in \mathbb{Z}}$, where θ is the left shift operator on the space of real valued bi-infinite sequences. Then, Λ is a compact metric space with the metric

$$d_\Lambda (\rho, \rho') := \sum_{n \in \mathbb{Z}} 2^{-|n|} |\rho_n - \lambda_n'|$$

and $\theta : \Lambda \to \Lambda$ is continuous, as the cocycle mapping defined by $\varphi(n, \rho, x_0) := x_n$.

The pullback attractor $A = \{A_\rho\}_{\rho \in \Lambda}$ for this skew product flow has component sets

$$A_\rho = \begin{cases} \{0\} & \text{for } \rho \in \Lambda \setminus \{\rho^+\}, \\ [1 - \lambda^+, \lambda^+ - 1] & \text{for } \rho = \rho^+. \end{cases}$$

The forward omega limit points end up in the set

$$\bigcup_{\rho \in \Lambda} A_\rho = \left[1 - \lambda^+, \lambda^+ - 1\right],$$

so the pullback attractor for the skew product flow under consideration contains all of the future asymptotic behaviour too. In contrast, the pullback attractor of the process for the difference equation (10.3) corresponding to the given sequence λ consists just of the zero solution.

Remark 10.4. The skew product formulation in this example provides more detailed information about the behaviour of the nonautonomous dynamical system than does the process formulation. Essentially, it represents a family of processes, one for each of the sequences in the base space Λ of the driving system. Moreover, since the base space Λ is compact, it also provides information about the future limiting behaviour too.

Chapter 11

Forward attractors

Autonomous systems depend only on the elapsed time, so their attractors and limit sets exist in current time. Similarly, the pullback limit defines an individual set of a nonautonomous pullback attractor at each instant of current time. The forward limit in the definition of a nonautonomous forward attractor is different as it is the limit to the asymptotically distant future. In particular, the limiting objects forward in time do not have the same dynamical meaning in current time as in the autonomous or pullback cases.

Until recently it was not known how to construct the component sets of a forward attractor. Here we show how this can be done, but unlike the pullback case the result is only a necessary condition. The sets so constructed must satisfy additional conditions to be forward attracting.

Remark 11.1. In this chapter and the following chapter we restrict attention to processes defined on the state space $X = \mathbb{R}^d$.

11.1 Pullback attractors recalled

We consider a continuous-time process ϕ on $\mathbb{R}^d$ and recall that a ϕ-invariant family $\mathcal{A} = \{A_t, t \in \mathbb{R}\}$ of nonempty compact subsets of $\mathbb{R}^d$ is called a *pullback attractor* if it pullback attracts all tempered families $\mathcal{D} = \{D_t, t \in \mathbb{R}\}$ of nonempty bounded subsets of $\mathbb{R}^d$, i.e.,

$$\lim_{t_0 \to -\infty} \text{dist}_{\mathbb{R}^d} \left(\phi(t, t_0, D_{t_0}), A_t \right) = 0, \qquad \text{(fixed } t\text{)}. \qquad (11.1)$$

Recall that the subsets D_t in a tempered family $\mathcal{D}$ cannot grow too quickly as $t \to \pm\infty$, See Definition 9.4.

The existence of a pullback attractor is ensured by that of a pullback absorbing family.

Theorem 11.1. *Suppose that a process ϕ on $\mathbb{R}^d$ has a ϕ-positively invariant pullback absorbing family $\mathcal{B} = \{B_t, t \in \mathbb{R}\}$ of nonempty compact subsets of $\mathbb{R}^d$, i.e., for each $t \in \mathbb{R}$ and every family $\mathcal{D} = \{D_t, t \in \mathbb{R}\}$ of nonempty tempered bounded subsets of $\mathbb{R}^d$ there exists a $T_{t,\mathcal{D}} \in \mathbb{R}^+$ such that*

$$\phi(t, t_0, D_{t_0}) \subseteq B_t \qquad \text{for all } t_0 \leq t - T_{t,\mathcal{D}}.$$

Then ϕ has a global pullback attractor $\mathcal{A} = \{A_t, t \in \mathbb{R}\}$ with component subsets determined by

$$A_t = \bigcap_{t_0 \leq t} \phi(t, t_0, B_{t_0}) \qquad \text{for each } t \in \mathbb{R}. \tag{11.2}$$

Moreover, if $\mathcal{A}$ is uniformly bounded, i.e., if $\bigcup_{t \in \mathbb{R}} A_t$ is bounded, then it is unique.

The uniqueness of the pullback attractor is assured if, for example, the pullback absorbing family $\mathcal{B}$ is uniformly bounded, i.e., if $\bigcup_{t \in \mathbb{R}} B_t$ is bounded.

Theorem 11.1 is essentially a necessary and sufficient condition for the existence of a pullback attractor, since it can be shown that every pullback attractor has a positively invariant pullback absorbing family.

11.2   Nonautonomous forward attractors in $\mathbb{R}^d$

Recall that a ϕ-invariant family $\mathcal{A} = \{A_t, t \in \mathbb{R}\}$ of nonempty compact subsets of $\mathbb{R}^d$ is called a *forward attractor* if it forward attracts all families $\mathcal{D} = \{D_t, t \in \mathbb{R}\}$ of nonempty tempered bounded subsets of $\mathbb{R}^d$, i.e.

$$\lim_{t \to \infty} \text{dist}_{\mathbb{R}^d}(\phi(t, t_0, D_{t_0}), A_t) = 0, \qquad \text{(fixed } t_0). \tag{11.3}$$

In contrast to the pullback limit in (11.1), the limit here involves a "moving" target.

Definition 11.1. A ϕ-invariant family $\mathcal{A} = \{A_t, t \in \mathbb{R}\}$ of nonempty compact subsets of $\mathbb{R}^d$ is said to be *Lyapunov stable* for a process ϕ if for each $t_0 \in \mathbb{R}$ and $\varepsilon > 0$ there exists $\delta = \delta(t_0, \varepsilon) > 0$ such that

$$\text{dist}_{\mathbb{R}^d}(\phi(t, t_0, x_0), A_t) < \varepsilon \tag{11.4}$$

for all $t \geq t_0$ and $x_0 \in \mathbb{R}^d$ with $\text{dist}_{\mathbb{R}^d}(x_0, A_{t_0}) \leq \delta$.

Proposition 11.1. *A forward attractor $\mathcal{A} = \{A_t, t \in \mathbb{R}\}$ of a process ϕ in $\mathbb{R}^d$ is Lyapunov asymptotically stable, i.e., both forward attracting in the sense of (11.3) and Lyapunov stable.*

This follows from the invariance of $\mathcal{A}$, the continuity with respect to the Hausdorff metric of the set-valued mapping $C \mapsto \phi(t, t_0, C)$ in compact sets uniformly on bounded time intervals $t \in [t_0, T]$ for each fixed t_0 and the forward attracting property (11.3). The proof is left to the reader. Note that it uses the compactness of the closed and bounded subset $B_\delta[A_{t_0}] = \{x \in \mathbb{R}^d : \text{dist}_{\mathbb{R}^d}(x_0, A_{t_0}) < \delta\}$ in $\mathbb{R}^d$.

Recall that a family $\mathcal{A} = \{A_t, t \in \mathbb{R}\}$ is uniformly bounded if there exists an $R > 0$ such that $A_t \subset B_R := \{x \in \mathbb{R}^d : \|x\| \leq R\}$ for each $t \in \mathbb{R}$ and define $B_1[A] := \{x \in \mathbb{R}^d : \text{dist}_{\mathbb{R}^d}(x, A) \leq 1\}$ for any nonempty compact subset A of $\mathbb{R}^d$.

Proposition 11.2. *A uniformly bounded forward attractor $\mathcal{A} = \{A_t, t \in \mathbb{R}\}$ of a process ϕ in $\mathbb{R}^d$ has a ϕ-positively invariant family $\mathcal{B} = \{B_t, t \in \mathbb{R}\}$ of nonempty compact subsets with $A_t \subset B_t$ for each $t \in \mathbb{R}$, which is forward absorbing.*

Proof. It can be assumed without loss of generality that R is large enough so that $B_1[A_t] \subset B_R$ for each $t \in \mathbb{R}$. By forward attraction of the set B_R, for each $t_0 \in \mathbb{R}$ there exists $T(t_0, 1) \geq 0$ such that

$$\text{dist}_{\mathbb{R}^d}\left(\phi(t, t_0, B_R), A_t\right) \leq 1, \qquad t \geq t_0 + T(t_0, 1).$$

Set $t_0 = 0$ and define t_1 to be the first integer with $t_1 \geq t_0 + T(t_0, 1)$. Then

$$\phi(t_1, t_0, B_1[A_{t_0}]) \subset \phi(t_1, t_0, B_R) \subset B_1[A_{t_1}].$$

Repeat this construction for $t_k < t_{k+1}$ for $k = 0, 1, 2, \ldots$ and define

$$B_t := \phi\left(t, t_k, B_1[A_{t_k}]\right), \qquad t_k \leq t < t_{k+1}, \ k = 0, 1, 2, \ldots.$$

These sets are obviously nonempty and compact. In particular, $B_{t_k} = B_1[A_{t_k}]$ and $\phi(t, t_0, B_{t_0}) \subset B_t$ for all $t \geq t_0 \geq 0$.

The construction of B_t proceeds differently for $t < 0$. Define $r_0 := 1$. Then the continuity with respect to the Hausdorff metric of the set-valued mapping $C \mapsto \phi(t, t_0, C)$ in compact sets uniformly on bounded time intervals $t \in [t_0, T]$ for each fixed t_0 with $\varepsilon = 1$, there exists $r_{-1} := \delta(-1, r_0) = \delta(-1, 1) > 0$ such that

$$\phi(0, -1, B_{r_{-1}}[A_{-1}] \cap B_R) \subset B_\varepsilon[A_0] = B_1[A_0] = B_1[A_0] \cap B_R.$$

Repeat this construction in the interval $[-k, -k+1]$ for $k = 1, 2, \ldots$ and define

$$B_t := \phi\left(t, -k, B_{r_{-k}}[A_{-k}] \cap B_R\right), \qquad -k \leq t < -k+1, \ k = 1, 2, \ldots$$

with radius $r_{-k} = \delta(-k, 1) > 0$ according to Proposition 11.1, respectively. These sets are obviously nonempty and compact with $B(-k) = B_{r_{-k}}[A_{-k}] \cap B_R$ and $\phi(t, t_0, B_{t_0}) \subset B_t$ for all $t_0 \leq t \leq 0$. Since $r_0 = 1$, the positive invariance property holds for all $t \in \mathbb{R}$.

The family $\mathcal{B} = \{B_t, t \in \mathbb{R}\}$ is by construction forward absorbing provided the time is taken large enough to reach one of the sets $B_1[A_{-k}]$ with $k \in \mathbb{N}$, which is possible by the forward attraction property. $\qquad\square$

11.3 Construction of possible forward attractors

It is possible to construct the component subsets of candidates for forward attractors with the same expression (11.2) as for a pullback attractor. This is based on the observation that a ϕ-positively invariant family of nonempty compact subsets of a process ϕ on $\mathbb{R}^d$ contains a maximal ϕ-invariant family of nonempty compact subsets. Essentially, it is formed by all the entire trajectories in the ϕ-positively invariant family.

Theorem 11.2. *Suppose that a process ϕ on $\mathbb{R}^d$ has a ϕ-positively invariant family $\mathcal{B} = \{B_t, t \in \mathbb{R}\}$ of nonempty compact subsets of $\mathbb{R}^d$. Then ϕ has a maximal ϕ-invariant family $\mathcal{A} = \{A_t, t \in \mathbb{R}\}$ in $\mathcal{B}$ of nonempty compact subsets determined by*

$$A_t = \bigcap_{t_0 \leq t} \phi\left(t, t_0, B_{t_0}\right) \qquad \text{for each } t \in \mathbb{R}. \tag{11.5}$$

In view of Proposition 11.2 the component sets of all forward attractors can be constructed in this way.

The proof of Theorem 11.2 is straightforward. The expression (11.5) is clear, since the sets $\{\phi(t, t_0, B_{t_0}), t_0 \le t\}$ with t fixed are nonempty and compact by the continuity of ϕ and the assumed compactness of the subsets B_{t_0}. They are also nested since by the two-parameter semigroup property and the positive invariance,

$$\phi(t, t_0, B_{t_0}) = \phi(t, t_1, \phi(t_1, t_0, B_{t_0})) \subset \phi(t, t_1, B_{t_1})$$

for (t, t_0), (t, t_1), $(t_1, t_0) \in \mathbb{R}_{\ge}^2 := \{(s, t) \in \mathbb{R}^2 : s \ge t\}$.

Remark 11.2. This pullback construction is used only inside the ϕ-positively invariant family $\mathcal{B}$. It is equivalent to

$$\lim_{t_0 \to -\infty} \mathrm{dist}_{\mathbb{R}^d}(\phi(t, t_0, B_{t_0}), A_t) = 0, \quad (\text{fixed } t). \tag{11.6}$$

Theorem 11.2 does not, however, imply that the subsets given by (11.5) form a pullback attractor, since nothing has been assumed so far about what is happening outside of the subsets B_t in $\mathcal{B}$. With the additional assumption that $\mathcal{B}$ is pullback absorbing, then Theorem 11.1 holds and the family $\mathcal{A} = \{A_t, t \in \mathbb{R}\}$ is a pullback attractor.

The situation with forward attractors is somewhat more complicated than for pullback attractors. This is a consequence of the different nature of the forward limit in (11.3) and the pullback limit in (11.1). For example, the forward attractor need not, however, be unique, even when it is uniformly bounded.

Example 11.1. Consider the nonautonomous scalar ODE

$$\dot{x} = -2tx \tag{11.7}$$

with explicit solutions $x(t, t_0, x_0) = x_0 e^{-t^2 + t_0^2}$.

For any $r \in \mathbb{R}^+$ the family $\mathcal{B}^{(r)} = \{B_t^{(r)}, t \in \mathbb{R}\}$ of nonempty compact subsets

$$B_t^{(r)} = \begin{cases} re^{-t^2}[-1, 1] : t \le 0, \\ r[-1, 1] \quad : t \ge 0, \end{cases}$$

is ϕ-positive invariant (and also forward absorbing). The corresponding subsets determined by (11.5) are $A_t^{(r)} = re^{-t^2}[-1, 1]$ for $t \in \mathbb{R}$.

Remark 11.3. The previous example shows that the forward attractor need not be unique since for any $r \in \mathbb{R}^+$, the sets $A_t^{(r)}$ are the component sets of a forward attractor $\mathcal{A}^{(r)}$. Interestingly, the unique solution of the ODE (11.7) for each initial value $x(t_0) = x_0$ is bounded and entire, i.e., defined for all $t \in \mathbb{R}$. In fact, every solution of this ODE is globally attracting in the forward sense and thus forms a forward attractor with component subsets consisting of singleton points of the solution.

Forward attractors are, however, *asymptotically equivalent.* Let $\mathcal{A}^{(1)} = \left\{A_t^{(1)}, t \in \mathbb{R}\right\}$ and $\mathcal{A}^{(2)} = \left\{A_t^{(2)}, t \in \mathbb{R}\right\}$ be two forward attractors with the same attraction universe. Then, by invariance and forward attraction,

$$\text{dist}_{\mathbb{R}^d}\left(A_t^{(2)}\, A_t^{(1)}\right) = \text{dist}_{\mathbb{R}^d}\left(\phi\left(t, t_0, A_{t_0}^{(2)}\right), A_t^{(1)}\right) \to 0 \quad \text{as} \ \to \infty$$

and similarly with the superscripts interchanged. Hence in the Hausdorff metric

$$H_{\mathbb{R}^d}\left(A_t^{(1)}\, A_t^{(2)}\right) \to 0 \quad \text{as} \ t \to \infty.$$

11.4 Forward asymptotic behaviour

The ϕ-invariant family $\mathcal{A} = \{A_t, t \in \mathbb{R}\}$ with subsets A_t given by the expression (11.5) in Theorem 11.2 does not necessarily reflect the full limiting dynamics in the forward time direction of the process. Recall the discussion in Chapter 10. In particular, a pullback attractor need not be forward attracting in the sense of the forward convergence in (11.3).

Example 11.2. Consider the process ϕ on $\mathbb{R}^1$ generated by the switching ODE

$$\dot{x} = f(t, x) := \begin{cases} -x & : t \leq 0, \\ x\left(1 - x^2\right) & : t > 0. \end{cases} \tag{11.8}$$

This nonautonomous system is asymptotically autonomous in both time directions with the limiting autonomous systems equal to the component systems holding on the whole time set $\mathbb{R}$.

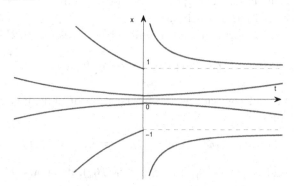

Fig. 11.1   Typical solutions of the switching system (11.8).

In the linear autonomous system the steady state solution 0 is asymptotically stable, while in the nonlinear autonomous system it is unstable and the steady state solutions ± 1 are (locally) asymptotically stable. These autonomous systems have global autonomous attractors given by $\{0\}$ and $[-1, 1]$, respectively.

The family $\mathcal{B}$ of constant sets $B_t = [-2, 2]$ is ϕ-positively invariant. The corresponding family $\mathcal{A}$ constructed by the pullback method in (11.5) has identical component subsets $A_t \equiv \{0\}$, $t \in \mathbb{R}$, corresponding to the zero entire solution and is the only bounded entire solution of the process.

Remark 11.4. Here $\mathcal{A} = \{A_t, t \in \mathbb{R}\}$ is, in fact, the pullback attractor of the nonautonomous system generated by the switching ODE (11.8), but that is not of interest to us here. More importantly, it is obviously not forward asymptotically attracting.

Indeed, the set $[-1, 1]$ of forward omega-limit points starting in the family $\mathcal{B}$ is not invariant for the process. (These are the limit points of convergent subsequences of sequences of points of the form $\phi(t_n, t_0, b_n)$, where $b_n \in B_{t_0}$ for some t_0 with $t_n \to \infty$, see below) In particular, the boundary points ± 1 are not entire solutions of the process, so cannot belong to a nonautonomous attractor, forward or pullback, since these consist of entire solutions.

To ensure that the family $\mathcal{A}$ constructed by the pullback method in (11.5) is forward attracting we need the set of omega limit points of trajectories in it to coincide with the set of omega limit points of trajectories starting in the family $\mathcal{B}$, rather than just to be a proper subset. In addition, the family $\mathcal{B}$ should be forward absorbing.

11.5 Omega limit points of a process

The asymptotic dynamics of the process ϕ on $\mathbb{R}^d$ inside the uniformly bounded ϕ-positively invariant family $\mathcal{B} = \{B_t, t \in \mathbb{R}\}$ of nonempty compact subsets B_t of $\mathbb{R}^d$ will be investigated more closely here.

For each $t_0 \in \mathbb{R}$, the forward omega limit set with respect to $\mathcal{B}$ is defined by

$$\omega_{\mathcal{B}, t_0} := \bigcap_{t \geq t_0} \overline{\bigcup_{s \geq t} \phi(s, t_0, B_{t_0})},$$

which is nonempty and compact as the intersection of nonempty nested compact subsets. In particular,

$$\lim_{t \to \infty} \mathrm{dist}_{\mathbb{R}^d} \left(\phi(t, t_0, B_{t_0}), \omega_{\mathcal{B}, t_0} \right) = 0, \quad \text{(fixed } t_0\text{)}. \tag{11.9}$$

Since $A_{t_0} \subset B_{t_0}$ and $A_t = \phi(t, t_0, A_{t_0}) \subset \phi(t, t_0, B_{t_0})$, it follows that

$$\lim_{t \to \infty} \mathrm{dist}_{\mathbb{R}^d} \left(A_t, \omega_{\mathcal{B}, t_0} \right) = 0, \quad \text{(fixed } t_0\text{)}. \tag{11.10}$$

Let B be the bounded subset of $\mathbb{R}$ such that $B_t \subset B$ for all $t \in \mathbb{R}$. Moreover, $\phi(t, t_0, B_{t_0}) \subset B_t \subset B$ for each $t \geq t_0$ since $\mathcal{B}$ is ϕ-positively invariant, so

$$\omega_{\mathcal{B}, t_0} \subset \omega_{\mathcal{B}, t_0'} \subset B, \qquad t_0 \leq t_0',$$

where the final inclusion is from the uniform boundedness of $\mathcal{B}$. Hence the sets

$$\omega_{\mathcal{B}}^{-\infty} := \bigcap_{t_0 \in \mathbb{R}} \omega_{\mathcal{B}, t_0}, \qquad \omega_{\mathcal{B}}^{+\infty} := \overline{\bigcup_{t_0 \in \mathbb{R}} \omega_{\mathcal{B}, t_0}}$$

are nonempty compact sets with $w_{\mathcal{B}}^{-\infty} \subset w_{\mathcal{B}}^{+\infty} \subset B$. From (11.9) it is clear that

$$\lim_{t\to\infty} \mathrm{dist}_{\mathbb{R}^d}\left(A_t, w_{\mathcal{B}}^{-\infty}\right) = 0. \tag{11.11}$$

Consider now the set of omega limit points for dynamics starting inside the family of sets $\mathcal{A} = \{A_t, t \in \mathbb{R}\}$ (note the sets here need not be nested), which is defined by

$$w_{\mathcal{A}} := \bigcap_{t_0 \in \mathbb{R}} \overline{\bigcup_{t \geq t_0} A_t} = \bigcap_{t_0 \in \mathbb{R}} \overline{\bigcup_{t \geq t_0} \phi(t, t_0, A_{t_0})}$$

which is nonempty and compact as a family of nested compact sets.

Obviously, $w_{\mathcal{A}} \subset w_{\mathcal{B}}^{-\infty} \subset w_{\mathcal{B}}^{+\infty} \subset B$. The inclusions here may be strict.

Example 11.3. For the process generated by the ODE (11.8), $w_{\mathcal{A}} = \{0\}$, while $w_{\mathcal{B}}^{-\infty} = w_{\mathcal{B}}^{+\infty} = [-1,1]$. The ODE (11.8) can be changed to include a second switching at, say, time $t = 10$ to the nonlinear ODE $\dot{x} = x(1-x^2)(x^2-4)$. Then $w_{\mathcal{B}}^{-\infty} = [-1,1]$, while $w_{\mathcal{B}}^{+\infty} = [-2,2]$.

11.6 Conditions ensuring forward convergence

The existence of omega limit points in $w_{\mathcal{B}}^{+\infty}$ that are not in $w_{\mathcal{A}}$ means that $\mathcal{A}$ cannot be forward attracting from within $\mathcal{B}$, i.e., starting at points $x_0 \in \mathbb{B}_{t_0}$ for $t_0 \in \mathbb{R}$. The converse also holds.

Theorem 11.3. $\mathcal{A}$ *is forward attracting from within* $\mathcal{B}$ *or a process* ϕ *on* $\mathbb{R}^d$, *i.e., the forward convergence in (11.3) holds, if and only if* $w_{\mathcal{A}} = w_{\mathcal{B}}^{+\infty}$.

Proof. Suppose that the forward convergence (11.3) does not hold. Then there is an $\varepsilon_0 > 0$ and a sequence $t_n \to \infty$ as $n \to \infty$ such that $\mathrm{dist}_{\mathbb{R}^d}\left(\phi(t_n, t_0, B_{t_0}), A_{t_n}\right) \geq \varepsilon_0$ for all $n \in \mathbb{N}$. Since the sets here are compact there exist points $\phi_n \in \phi(t_n, t_0, B_{t_0})$ such that

$$\mathrm{dist}_{\mathbb{R}^d}\left(\phi_n, A_{t_n}\right) = \mathrm{dist}_{\mathbb{R}^d}\left(\phi(t_n, t_0, B_{t_0}), A_{t_n}\right) \geq \varepsilon_0.$$

Then for any points $a_n \in A_{t_n}$,

$$\|\phi_n - a_n\| \geq \mathrm{dist}_{\mathbb{R}^d}\left(\phi_n, A_{t_n}\right) \geq \varepsilon_0.$$

In particular, $\|\phi_n - a_n\| \geq \varepsilon_0$ for all $n \in \mathbb{N}$.

Now the ϕ_n and a_n belong to the compact set B, so there are convergent subsequences $\phi_{n_j} \to \bar{\phi} \in B$ and $a_{n_j} \to \bar{a} \in B$. It follows that $\|\bar{\phi} - \bar{a}\| \geq \varepsilon_0$. From the definitions $\bar{\phi} \in w_{\mathcal{B},t_0} \subset w_{\mathcal{B}}^{+\infty}$ and $\bar{a} \in w_{\mathcal{A}}$. Since the a_n and hence $\bar{a}$ were otherwise arbitrary, it follows that $\mathrm{dist}_{\mathbb{R}^d}\left(\bar{\phi}, w_{\mathcal{A}}\right) \geq \varepsilon_0$, so $\mathrm{dist}_{\mathbb{R}^d}\left(w_{\mathcal{B}}^{+\infty}, w_{\mathcal{A}}\right) \geq \varepsilon_0$ and hence $w_{\mathcal{A}} \neq w_{\mathcal{B}}^{+\infty}$. $\qquad\square$

Forward attraction also holds when the rate of pullback convergence from within $\mathcal{B}$ to construct the component sets A_t of $\mathcal{A}$ in Theorem 11.2 is uniform. In fact, it suffices that the convergence rate is eventually uniform.

Theorem 11.4. *For a process ϕ on $\mathbb{R}^d$ the family $\mathcal{A} = \{A_t, t \in \mathbb{R}\}$ with subsets A_t constructed by (11.5) is forward attracting from within $\mathcal{B}$, i.e., the forward convergence (11.3) holds, if the rate of pullback convergence from within $\mathcal{B}$ to the component sets A_t of $\mathcal{A}$ is eventually uniform in the sense that for every $\varepsilon > 0$ there exist $\tau(\varepsilon) \in \mathbb{R}$ and $T(\varepsilon) > 0$ such that for each $t \geq \tau(\varepsilon)$*

$$\mathrm{dist}_{\mathbb{R}^d}\left(\phi(t, t_0, B_{t_0}), A_t\right) < \varepsilon \qquad (11.12)$$

holds for all $t_0 \leq t - T(\varepsilon)$.

Proof. Given any $t_0 \in \mathbb{R}$ and $\varepsilon > 0$, it is clear that the inequality (11.12) holds for every $t \geq \max\{t_0, \tau(\varepsilon)\} + T(\varepsilon)$, i.e., the forward convergence (11.3) holds. $\qquad \square$

Theorem 11.4 is concerned only with the rate of pullback convergence starting from inside the positively invariant family $\mathcal{B}$. It makes no assumptions about what is happening outside of $\mathcal{B}$ such as pullback or forward absorption.

Remark 11.5. The family $\mathcal{A}$ for the simple switching system generated by the ODE (11.8) is not forward attracting. By considering any time $t > 0$, it is not hard to see that the rate of pullback convergence is not eventually uniform.

11.7   Asymptotically autonomous systems

The ODE (11.8) generates a simple process ϕ on $\mathbb{R}^1$ formed by switching between two autonomous systems. It is a special case of an asymptotically autonomous system, see, e.g., Hale [Hale (1988)], Kato *et al.* [Kato, Martynyuk and Sheshakov (1996)] and Lasalle [Lasalle (1976)]. These have additional structure that provides an easier way to check the condition of Theorem 11.3 for forward convergence. theorem in [Kloeden and Simsen (2015)].

Theorem 11.5. *Let ϕ be a uniformly bounded process on $\mathbb{R}^d$ with a ϕ-positively invariant family $\mathcal{B} = \{B_t, t \in \mathbb{R}\}$ of nonempty compact subsets of $\mathbb{R}^d$ with all component subsets contained in a compact set B. Let π be an autonomous semi-dynamical system on $\mathbb{R}^d$ with a global attractor A_∞ in B. In addition, suppose that*

$$\phi(t + \tau, \tau, x_\tau) \to \pi(t, x_0) \quad \text{as } \tau \to +\infty \qquad (11.13)$$

uniformly in $t \in \mathbb{R}^+$ whenever $x_\tau \in B_\tau$ and $x_\tau \to x_0$ as $\tau \to +\infty$.
Then $\omega_{\mathcal{B}, t_0}^{+\infty} \subset A_\infty$ for each $t_0 \in \mathbb{R}$ and

$$\lim_{t \to +\infty} \mathrm{dist}_{\mathbb{R}^d}\left(A_t, A_\infty\right) = 0, \qquad (11.14)$$

where $\mathcal{A} = \{A_t, t \in \mathbb{R}\}$ is the family of nonempty compact subsets of $\mathbb{R}^d$ with all component subsets defined by pullback characterisation (11.5) from within $\mathcal{B}$.

Proof. Fix $t \in \mathbb{R}^+$ and $t_0 \in \mathbb{R}$ and define $s_n := t_n - t$, where the sequence $t_n \to \infty$ as $n \to \infty$. By the 2-parameter semigroup property and the ϕ-positive invariance of $\mathcal{B}$

$$\phi(t_n, t_0, B_{t_0}) = \phi(t_n, s_n, \phi(s_n, t_0, B_{t_0})) \subset \phi(t_n, s_n, B(s_n)).$$

Hence $\phi(t + s_n, t_0, B_{t_0}) \subset \phi(t + s_n, s_n, B_{s_n})$. By compactness of the sets involved, for each $n \in \mathbb{N}$, there exists a $b_n \in B_{s_n} \subset B$ such that

$$\mathrm{dist}_{\mathbb{R}^d}(\phi(t + s_n, s_n, b_n), A_\infty) = \mathrm{dist}_{\mathbb{R}^d}(\phi(t + s_n, s_n, B_{s_n}), A_\infty).$$

By the compactness of the set B there is a convergent subsequence (labelled for convenience as the original one) $b_n \to \bar{b} \in B$ as $n \to \infty$. Hence,

$$\mathrm{dist}_{\mathbb{R}^d}(\phi(t + s_n, s_n, b_n), A_\infty) \leq \mathrm{dist}_{\mathbb{R}^d}(\phi(t + s_n, s_n, b_n), \pi(t, \bar{b}))$$

$$+ \mathrm{dist}_{\mathbb{R}^d}(\pi(t, \bar{b}), A_\infty)$$

$$\leq \varepsilon + \mathrm{dist}_{\mathbb{R}^d}(\pi(t, \bar{b}), A_\infty)$$

for any $\varepsilon > 0$ and $n \geq N(\varepsilon)$ uniformly in $t \in \mathbb{R}$ by the asymptotic autonomous condition (11.13). Since A_∞ is the global attractor of π for any $\varepsilon > 0$ there exists a $T(\varepsilon, B) \geq 0$ such that

$$\mathrm{dist}_{\mathbb{R}^d}(\pi(t, \bar{b}), A_\infty) \leq \mathrm{dist}_{\mathbb{R}^d}(\pi(t, B), A_\infty) < \varepsilon \quad \text{for all } t \geq T(\varepsilon, B).$$

Combining the results gives

$$\lim_{t \to +\infty} \mathrm{dist}_{\mathbb{R}^d}(\phi(t, t_0, B_{t_0}), A_\infty) = 0, \tag{11.15}$$

for each $t_0 \in \mathbb{R}$. This means that $\omega_{B,t_0}^{+\infty} \subset A_\infty$ for each $t_0 \in \mathbb{R}$.

The limit (11.14) then follows since $A_t = \phi(t, t_0, A_{t_0}) \subset \phi(t, t_0, B_{t_0})$, so

$$\mathrm{dist}_{\mathbb{R}^d}(A_t, A_\infty) = \mathrm{dist}_{\mathbb{R}^d}(\phi(t, t_0, A_{t_0}), A_\infty) \leq \mathrm{dist}_{\mathbb{R}^d}(\phi(t, t_0, B_{t_0}), A_\infty).$$

$\square$

Combining Theorem 11.3 and the limit (11.15) in the proof of Theorem 11.5 gives:

Corollary 11.1. *Suppose that the assumptions of Theorems 11.3 and 11.5 hold and that $\omega_A = A_\infty$. Then $A = \{A_t, t \in \mathbb{R}\}$ is forward attracting from within $\mathcal{B}$, i.e., the forward convergence in (11.3) holds.*

Condition 11.13 in Theorem 11.5 can be easily checked directly in terms of the continuous mappings f and f_n of the autonomous and nonautonomous differences equation

$$x_{n+1} = f(x_n), \qquad x_{n+1} = f_n(x_n), \qquad n \in \mathbb{Z}$$

provided the compact set B in Theorem 11.5 is positively invariant under f and the f_n, i.e., $f(B) \subset B$ and $f_n(B) \subset B$, for all large n.

Let ϕ be the process generated by the nonautonomous difference equation and π be the semi-group generated by the autonomous difference equation.

Lemma 11.1. *Condition (11.13) of Theorem 11.5 holds if*

$$\lim_{n \to \infty} \text{dist}_{\mathbb{R}^d} \big(f_n(x_n), f(x_0) \big) = 0 \qquad \textit{for all } x_n \to x_0, \tag{11.16}$$

provided the compact set B in Theorem 11.5 is positively invariant under f and the f_n, i.e., $f(B) \subset B$ and $f_n(B) \subset B$ for all large n.

Proof. The proof is by induction. Condition (11.13) is true for $N = 1$ since it then reduces to (11.16).

Suppose it is true for $N \geq 1$ and consider the case with $N + 1$. By the two-parameter semi-group property

$$\phi(n + N + 1, n, x_n) = \phi(n + N + 1, n + N, \varphi(n + N, n, x_n)) = f_{n+N}(\varphi(n + N, n, x_n))$$

and by the semi-group property

$$\pi(N + 1, x_0) = \pi(1, \pi(N, x_0)) = f(\pi(N, x_0)).$$

Hence

$$\phi(n + N + 1, n, x_n) - \pi(N + 1, x_0) = f_{n+N}(\varphi(n + N, n, x_n)) - f(\pi(N, x_0))$$
$$= f_{n+N}(z_{n+N}) - f(z),$$

where

$$z_{n+N} := \phi(n + N, n, x_n) \to z := \pi(N, x_0)$$

by the induction assumption. Relabelling the sequences with an index $k = n + N$, it follows that

$$\phi(n + N + 1, n, x_n) \to \pi(N + 1, x_0),$$

so the result is then true for $N + 1$. □

Chapter 12

Omega-limit sets and forward attracting sets

In autonomous dynamical systems the attractor corresponds to the ω-limit set ω_B of an absorbing set B. Although this set is defined as the limit as time goes to infinity, it actually exists for all time, since the time variable in an autonomous dynamical system $\pi(t, x)$ is just the elapsed time from the start and not the actual time. In particular, the ω-limit set ω_B here is an invariant set.

The situation is rather different in a nonautonomous dynamical system described by a process $\phi(t, t_0, x)$ which depends on both the actual time and starting time and not only their difference. ω-limit sets can be defined here, but depend on the starting time, i.e., ω_{B,t_0} [Lasalle (1976)]. These are increasing in t_0 and the closure of their union ω_B represents the set of all future limit points of the system.

Haraux [Haraux (1991)] introduced the concept of a *uniform attractor*[1] as the smallest set that attracts all bounded sets uniformly in the initial time. This was extensively investigated in applications by Vishik and his coworkers [Chepyzhov and Vishik (2002); Vishik (1992)] It contains all forward ω-limit points, hence ω_B, but can be a larger set as an example in the appendix of [Carvalho, Langa and Robinson (2013)] shows.

However, neither the Haraux-Vishik uniform attractor nor ω_B need be invariant or even positive invariant. Nevertheless, as will be shown in this chapter, they are nearly invariant in an asymptotic sense, i.e., asymptotically positive invariant and also asymptotically negative invariant under some increasingly stronger uniformity assumptions on the future dynamics. They thus resemble attractors as usually understood, i.e., with the properties of attraction, compactness and invariance.

We saw in earlier chapters that skew product flows with a driving system on a compact base space allow much more to be said about the future asymptotic dynamics of the system. These are important, but special cases, which include many interesting situations.

---

[1] This is not to be confused with the uniform (nonautonomous) attractor introduced in earlier, which is uniformly both a forward and pullback attractor.

12.1 Forward attracting sets

Let ϕ be a continuous time process on $\mathbb{R}^d$ and suppose that ϕ is dissipative, i.e., it has a closed and bounded (hence compact) absorbing set B in $\mathbb{R}^d$, which is eventually uniformly absorbing.

Assumption 12.1. There exists a ϕ-positive invariant compact subset B in $\mathbb{R}^d$ for a process ϕ on $\mathbb{R}^d$ and a $T^* \geq 0$ such that for any bounded subset D of $\mathbb{R}^d$ and every $t_0 \geq T^*$ there exists a $T_D \geq 0$ for which

$$\phi(t, t_0, x_0) \in B \quad \forall t \geq t_0 + T_D, x_0 \in D.$$

It follows from this dissipativity assumption and the compactness of the set B that the ω-limit set

$$\omega_{B,t_0} := \bigcap_{t \geq t_0} \overline{\bigcup_{s \geq t} \phi(s, t_0, B)}$$

is a nonempty compact set of $\mathbb{R}^d$ for each $t_0 \geq T^*$. Note that

$$\lim_{t \to \infty} \mathrm{dist}_{\mathbb{R}^d} \left(\phi(t, t_0, B), \omega_{B,t_0} \right) = 0 \tag{12.1}$$

for each $t_0 \geq T^*$ and that $\omega_{B,t_0} \subset \omega_{B,t_0'} \subset B$ for $t_0 \leq t_0'$. Hence, the set

$$\omega_B := \overline{\bigcup_{t_0 \geq T^*} \omega_{B,t_0}} \subset B$$

is nonempty and compact. It contains all of the future limit points of the process starting in the set B at some time $t_0 \geq T^*$. In particular, $\bar{y} \in \omega_B$ if there exist sequences $b_{0,j} \in B$ and $t_{0,j} \leq t_{0,j}'$ with $t_{0,j} \to \infty$ as $j \to \infty$ such that $\phi(t_{0,j}', t_{0,j}, b_{0,j}) \to \bar{y}$ as $j \to \infty$.

The set ω_B characterises the forward asymptotic behaviour of the nonautonomous system. It was called the *forward attracting set* of the nonautonomous system in [Kloeden and Yang (2016)] and is closely related to the Haraux-Vishik *uniform attractor*, but it may be smaller and does not require the generating ODE (12.5) to be defined for all time or the attraction to be uniform in the initial time [Kloeden and Lorenz (2016); Kloeden and Yang (2016)].

Example 12.1. The scalar ODE $\dot{x} = -x + e^{-t}$ with $\omega_B = \{0\}$ shows that the set ω_B need not be invariant or even positive invariant. See Figure 12.1.

12.1.1 *Asymptotically positive invariance*

The set $\omega_B = \{0\}$ in the previous example appears to become more and more invariant the later one starts in the future.

This motivated the concept of asymptotically positive invariance in the literature for nonautonomous differential equations [Kloeden (1975); Lakshmikantham and Leela (1967)]. Positive invariance says that a set is mapped into itself at future

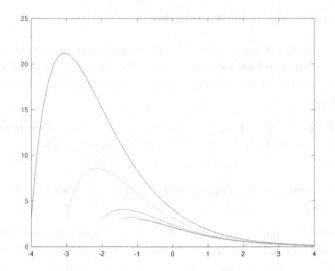

Fig. 12.1   Solutions of $\dot{x} = -x + e^{-t}$ with different initial conditions.

times, while asymptotical positive invariance says it is mapped almost into itself and closer the later one starts.

Definition 12.1. A set A is said to be *asymptotically positive invariant* for a process ϕ on $\mathbb{R}^d$ if for every $\varepsilon > 0$ here exists a $T(\varepsilon)$ such that

$$\phi(t, t_0, A) \subset B_\varepsilon(A), \quad t \geq t_0, \tag{12.2}$$

for each $t_0 \geq T(\varepsilon)$, where $B_\varepsilon(A) := \{x \in \mathbb{R}^d : \text{dist}_{\mathbb{R}^d}(x, A) < \varepsilon\}$.

Theorem 12.1. *Suppose that Assumption 12.1 hold for a process ϕ on $\mathbb{R}^d$. Then ω_B is asymptotically positive invariant.*

Proof. Since $\omega_{B,t_0} \subset \omega_B \subset B$ and $\phi(t, t_0, \omega_B) \subset \phi(t, t_0, B) \subset B$, by the convergence (12.1), for every $\varepsilon > 0$ and $t_0 \geq T^*$ there exists $T_0(t_0, \varepsilon) \in \mathbb{R}^+$ such that

$$\text{dist}_{\mathbb{R}^d}(\phi(t, t_0, \omega_B), \omega_B) < \varepsilon \quad \text{for } t \geq t_0 + T_0(t_0, \varepsilon).$$

Suppose that for some $\varepsilon_0 > 0$ there are sequences $t_{0,j} \leq t_j \leq t_{0,j} + T_0(t_{0,j}, \varepsilon_0)$ with $t_{0,j} \to \infty$ as $j \to \infty$ such that

$$\text{dist}_{\mathbb{R}^d}(\phi(t_j, t_{0,j}, \omega_B), \omega_B) \geq \varepsilon_0, \quad j \in \mathbb{N}.$$

Since $\phi(t_j, t_{0,j}, \omega_B)$ is compact there exists a $b_j \in \omega_B \subset B$ such that

$$\text{dist}_{\mathbb{R}^d}(\phi(t_j, t_{0,j}, b_j), \omega_B) = \text{dist}_{\mathbb{R}^d}(\phi(t_j, t_{0,j}, \omega_B), \omega_B) \geq \varepsilon_0, \quad j \in \mathbb{N}.$$

Define $y_j := \phi(t_j, t_{0,j}, b_j)$. Since the points $y_j \in B$, which is compact, there exists a convergent subsequence $y_{j_k} \to \bar{y} \in B$. Moreover, $\bar{y} \in \omega_B$ by the definition. However, $\text{dist}_{\mathbb{R}^d}(y_j, \omega_B) \geq \varepsilon_0$, so $\text{dist}_{\mathbb{R}^d}(\bar{y}, \omega_B) \geq \varepsilon_0$, which is a contradiction.

Hence for any $\varepsilon > 0$ there exists a $T(\varepsilon)$ large enough such that

$$\text{dist}_{\mathbb{R}^d}(\phi(t, t_0, \omega_B), \omega_B) < \varepsilon \quad \text{for } t \geq t_0 \geq T\varepsilon).$$

It follows that ω_B is asympotically positive invariant for the process ϕ on $\mathbb{R}^d$. $\square$

12.2   Asymptotically negative invariance

The concept of negative invariance of a set implies that any point in it can be reached in any even time from another point in it. The set ω_B is generally not negative invariant, but under an additional uniformity assumption it is asymptotically negative invariant.

Definition 12.2. A set A is said to be *asymptotically negative invariant* for a process ϕ if for every $a \in A$, $\varepsilon > 0$ and $T > 0$, there exist t_ε and $a_\varepsilon \in A$ such that
$$\|\phi\left(t_\varepsilon, t_\varepsilon - T, a_\varepsilon\right) - a\| < \varepsilon.$$

To show that this property holds a further assumption about the future uniform behaviour of the process is needed.

Assumption 12.2. There exists a $T^* \geq 0$ such that the process ϕ on $\mathbb{R}^d$ is Lipschitz continuous in initial conditions in B on finite time intervals $[t_0, t_0 + T]$ uniformly in $t_0 \geq T^*$, i.e., there exists a constant $L_B > 0$ independent of $t_0 \geq T^*$ such that
$$\|\phi\left(t, \tau, x_0\right) - \phi\left(t, \tau, y_0\right)\| \leq \|x_0 - y_0\| e^{L_B(t-\tau)} \leq \|x_0 - y_0\| e^{L_B T} \tag{12.3}$$
for all $x_0, y_0 \in B$ and $T^* \leq \tau < t \leq \tau + T$.

Theorem 12.2. *Suppose that the Assumptions 12.1 and 12.2 hold for a process ϕ on $\mathbb{R}^d$. Then ω_B is asymptotically negative invariant.*

Proof. To show this let $\omega \in \omega_B$, $\varepsilon > 0$ and $T > 0$ be given. Then there exist sequences $b_n \in B$ and $\tau_n < t_n$ with $\tau_n \to \infty$ and an integer $N(\varepsilon)$ such that
$$\|\phi\left(t_n, \tau_n, b_n\right) - \omega\| < \frac{1}{2}\varepsilon, \qquad n \geq N(\varepsilon).$$
Define $a_n := \phi\left(t_n - T, \tau_n, b_n\right) \in B$. Since B is compact, there exists a convergent subsequence $a_{n_j} := \phi\left(t_{n_j} - T, \tau_{n_j}, b_{n_j}\right) \to \omega_\varepsilon$ as $n_j \to \infty$. By definition, $\omega_\varepsilon \in \omega_B$.

From Assumption 12.2 the process ϕ is continuous in initial conditions uniformly on finite time intervals of the same length, i.e., (12.3). Hence
$$\left\|\phi\left(t_{n_j}, t_{n_j} - T, a_{n_j}\right) - \phi\left(t_{n_j}, t_{n_j} - T, \omega_\varepsilon\right)\right\| < \frac{1}{2}\varepsilon, \qquad n_j \geq \widehat{N}(\varepsilon). \tag{12.4}$$
By the 2-parameter semi-group property
$$\phi\left(t_{n_j}, t_{n_j} - T, a_{n_j}\right) = \phi\left(t_{n_j}, t_{n_j} - T, \phi\left(t_{n_j}, \tau_{n_j}, b_{n_j}\right)\right) = \phi\left(t_{n_j}, \tau_{n_j}, b_{n_j}\right).$$
Then
$$\left\|\omega - \phi\left(t_{n_j}, t_{n_j} - T, \omega_\varepsilon\right)\right\| \leq \left\|\omega - \phi\left(t_{n_j}, t_{n_j} - T, a_{n_j}\right)\right\|$$
$$+ \left\|\phi\left(t_{n_j}, t_{n_j} - T, a_{n_j}\right) - \phi\left(t_{n_j}, t_{n_j} - T, \omega_\varepsilon\right)\right\|$$
$$\leq \left\|\omega - \phi\left(t_{n_j}, \tau_{n_j}, b_{n_j}\right)\right\|$$
$$+ \left\|\phi\left(t_{n_j}, t_{n_j} - T, a_{n_j}\right) - \phi\left(t_{n_j}, t_{n_j} - T, \omega_\varepsilon\right)\right\|$$
$$< \frac{1}{2}\varepsilon + \frac{1}{2}\varepsilon = \varepsilon.$$

This is the desired result. $\qquad\square$

Example 12.2. Consider a nonautonomous ODE in $\mathbb{R}^d$

$$\frac{dx}{dt} = f(t, x) \quad \text{for all } t \geq T^*, x \in \mathbb{R}^d, \tag{12.5}$$

where $f : [T^*, \infty) \times \mathbb{R}^d \to \mathbb{R}^d$ is (at least) continuously differentiable, so there exists a unique solution $\phi(t, t_0, x_0)$ for the initial value $x(t_0) = x_0$.

Assume that there is a compact positive invariant absorbing set B as in Assumption 12.1 and that the mappings $t \mapsto f(t, x)$ and $t \mapsto \nabla_x f(t, x)$ are uniformly continuous in $t \geq T^*$ uniformly in $x \in B$.

Hence the vector field f of the ODE (12.5) is Lipschitz continuous on the compact absorbing set B uniformly in time, i.e.,

$$\|f(t, x) - f(t, y)\| \leq L_B \|x - y\|, \qquad t \geq T^*, x, y \in B, \tag{12.6}$$

with the constant $L_B = \sup_{x \in B, t \geq T^*} \|\nabla_x f(t, x)\| < \infty$.

Thus for any two solutions $\phi(t, \tau, x_0)$, $\phi(t, \tau, y_0)$ in B

$$\|\phi(t, \tau, x_0) - \phi(t, \tau, y_0)\|$$

$$\leq \|x_0 - y_0\| + \int_\tau^t \|f(sm\phi(s, \tau, x_0)) - f(s, \phi(s, \tau, y_0))\| \, ds$$

$$\leq \|x_0 - y_0\| + L_B \int_\tau^t \|\phi(s, \tau, x_0) - \phi(s, \tau, y_0)\| \, ds.$$

Gronwall's inequality then gives

$$\|\phi(t, \tau, x_0) - \phi(t, \tau, y_0)\| \leq \|x_0 - y_0\| e^{L_B(t-\tau)} \leq \|x_0 - y_0\| e^{L_B T}, \tag{12.7}$$

where $0 \leq t - \tau \leq T$. (Note that the bound depends just on the length of the time interval and not on the starting point of the interval).

Thus the process generated by the ODE (12.5) satisfies Assumption 12.2.

Remark 12.1. The uniformly continuity assumption on the mappings $t \mapsto f(t, x)$ and $t \mapsto \nabla_x f(t, x)$ n Example 12.2 hold if, for example, f has the form $f(t, x) = \hat{f}(x, \psi(t))$, where $\psi : [T^*, \infty) \to \mathbb{R}^m$ is uniformly bounded and uniformly continuous, such as an almost periodic or recurrent function, or a more general function as in [Kloeden and Rodrigues (2011)].

Example 12.3. Consider a nonautonomous difference equation

$$x_{n+1} = f_n(x_n) \quad \text{for } N \geq N^*, \tag{12.8}$$

where the $f_n : \mathbb{R}^d \to \mathbb{R}^d$ are continuous. This generates a discrete time process

$$x_n = \phi(n, n_0, x_0) := f_{n-1} \circ \cdots \circ f_{n_0}(x_0) \quad \text{for all } x_0 \in \mathbb{R}^d, n \geq n_0 \geq T^*.$$

Suppose that ϕ satisfies Assumption 12.1 and Assumption 12.2, i.e., has a compact positively invariant absorbing set $B \subset \mathbb{R}^d$ and is uniformly Lipschitz continuous in initial conditions in B on finite time sets $[n_0, n_0 + K] \cap \mathbb{Z}_{\geq N^*}$ for any $K \in \mathbb{N}$ independently of the choice $n_0 \geq N^*$.

Note that Assumption 12.2 holds of the f_n are Lipschitz on B uniformly in $n \geq N^*$, i.e., there exists a constant L such that

$$\|f_n(x_0) - f_n(y_0)\| \leq L \|x_0 - y_0\|, \quad x_0, y_0 \in B, n \geq N^*,$$

since $f_k(B) \subset B$ for all $k \geq N^*$ and hence $f_{n-1} \circ \cdots \circ f_{n_0}(B) \subset B$. Thus for any $x_0, y_0 \in B$ by induction

$$\|f_{n-1} \circ \cdots \circ f_{n_0}(x_0) - f_{n-1} \circ \cdots \circ f_{n_0}(y_0)\|$$

$$= \|f_{n-1}(f_{n-2} \circ \cdots \circ f_{n_0}(x_0)) - f_{n-1}(f_{n-2} \circ \cdots \circ f_{n_0}(y_0))\|$$

$$\leq L \|f_{n-2} \circ \cdots \circ f_{n_0}(x_0) - f_{n-2} \circ \cdots \circ f_{n_0}(y_0)\|$$

$$\vdots \qquad \vdots \qquad \vdots$$

$$\leq L^{n-n_0} \|x_0 - y_0\|$$

for $n = n_0 + 1, \cdots, n_0 + N$, so

$$\|\phi(n_0+N, n_0, x_0) - \phi(n_0+N, n_0, y_0)\| \leq L^N \|x_0 - y_0\|, \quad x_0, y_0 \in B, n_0 \geq N^*, N \geq 1.$$

12.3   Upper semi-continuity in a parameter

A classical result in the theory of autonomous dynamical systems is that the attractor depend upper semi-continuously on a parameter. Now consider a parameterised family of processes $\phi^\nu(t, t_0, x_0)$, where $\nu \in [0, \nu^*]$, on $\mathbb{R}^d$. A similar results will be established here for the omega limit sets of a a parametrised family of processes under appropriate future uniformity assumptions.

For this Assumption 12.1 is strengthened so that the parameterised family of processes ϕ^ν on $\mathbb{R}^d$ is equi-absorbing in a compact absorbing B uniformly in the parameter $\nu \in [0, \nu^*]$.

Assumption 12.3. The exists nonempty compact set B which is ϕ^ν-positive invariant for each $\nu \in [0, \nu^*]$ and for any bounded subset D of $\mathbb{R}^d$ and $t_0 \geq T^*$ there exists a $T_D \geq 0$ (independent of ν) such that

$$\phi^\nu(t, t_0, x_0) \in B \quad \text{for all } t \geq t_0 + T_D, x_0 \in D, \nu \in [0, \nu^*].$$

In addition, the following uniform continuous convergence is assumed.

Assumption 12.4. For every $\varepsilon > 0$ and $T > 0$ there exists a $\delta(\varepsilon, T) > 0$ such that

$$\|\phi^\nu(t, t_0, b)) - \phi^0(t, t_0, b))\| < \varepsilon, \quad t_0 \leq t \leq t_0 + T, b \in B.$$

for $|\nu| < \delta(\varepsilon, T)$ and $t_0 \geq T^*$.

Finally, the uniform attraction of the set ω_B^0 for the system ϕ_0 is also needed for the following result.

Assumption 12.5. ω_B^0 uniformly attracts the set B under the process ϕ^0 on $\mathbb{R}^d$, i.e., for every $\varepsilon > 0$ there exists a $T(\varepsilon)$, which is independent of $t_0 \geq T^*$, such that

$$\text{dist}_{\mathbb{R}^d}\left(\phi^0(t_0 + t, t_0, B), \omega_B^0\right) < \varepsilon, \quad t \geq T(\varepsilon), t_0 \geq T^*.$$

Then the upper semi-continuous convergence of the forward attracting sets in the parameter holds.

Theorem 12.3. *Suppose that Assumptions 12.3, 12.4 and 12.5 hold for the family of processes ϕ^ν on $\mathbb{R}^d$. Then*

$$\text{dist}_{\mathbb{R}^d}\left(\omega_B^\nu, \omega_B^0\right) \to 0 \qquad \text{as } \nu \to 0.$$

Proof. A proof by contradiction will be used.

Suppose for some sequence of parameters $\nu_j \to 0$ that the above limit is not true, i.e., there exists an $\varepsilon_0 > 0$ such that

$$\text{dist}_{\mathbb{R}^d}\left(\omega_B^{\nu_j}, \omega_B^0\right) \geq \varepsilon_0, \quad j \in \mathbb{N}.$$

Since $\omega_B^{\nu_j}$ is compact, there exists $\omega_j \in \omega_B^{\nu_j}$ such that

$$\text{dist}_{\mathbb{R}^d}\left(\omega_j, \omega_B^0\right) = \text{dist}_{\mathbb{R}^d}\left(\omega_B^{\nu_j}, \omega_B^0\right) \geq \varepsilon_0, \quad j \in \mathbb{N}. \tag{12.9}$$

By Assumption 12.5 there is a $T = T(\varepsilon_0/8)$ such that for any $t_0 \geq T^*$

$$\text{dist}_{\mathbb{R}^d}\left(\phi^0\left(t_0 + T, t_0, B\right), \omega_B^0\right) < \frac{1}{8}\varepsilon_0.$$

Then use Assumption 12.4 with this T to pick a $\nu_j < \delta(\varepsilon_0/2, T)$ to ensure that

$$\left\|\phi^{\nu_j}\left(t_0, t_0 - T, b\right)\right) - \phi^0\left(t, t_0 - T, b\right)\right\| < \frac{1}{2}\varepsilon_0, \qquad b \in B, t_0 \gg 0. \tag{12.10}$$

Fix such a ν_j and use the asymptotical negative invariance of $\omega_B^{\nu_j}$ to obtain the existence of an $\omega_{j,T} \in \omega_B^{\nu_j}$ and a $t_\varepsilon^j \gg 0$ so that

$$\left\|\phi^{\nu_j}\left(t_\varepsilon^j, t_\varepsilon^j - T, \omega_{j,T}\right) - \omega_j\right\| < \frac{1}{8}\varepsilon_0.$$

Then, with t_0 taken as t_ε^j above,

$$\begin{aligned}
\text{dist}_{\mathbb{R}^d}\left(\omega_j, \omega_B^0\right) &\leq \left\|\omega_j - \phi^{\nu_j}\left(t_\varepsilon^j, t_\varepsilon^j - T, \omega_{j,T}\right)\right\| \\
&\quad + \left\|\phi^{\nu_j}\left(t_\varepsilon^j, t_\varepsilon^j - T, \omega_{j,T}\right) - \phi^0\left(t_\varepsilon^j, t_\varepsilon^j - T, \omega_{j,T}\right)\right\| \\
&\quad + \text{dist}_{\mathbb{R}^d}\left(\phi^0\left(t_\varepsilon^j, t_\varepsilon^j - T, \omega_{j,T}\right), \omega_B^0\right) \\
&< \frac{1}{8}\varepsilon_0 + \frac{1}{2}\varepsilon_0 + \frac{1}{8}\varepsilon_0 = \frac{3}{4}\varepsilon_0,
\end{aligned}$$

which contradicts the assumption (12.9). $\qquad\square$

Example 12.4. Consider a parameterised family of nonautonomous ODEs in $\mathbb{R}^d$

$$\frac{dx}{dt} = f^\nu(t, x) \quad \text{for all } t \geq T^*, \; x \in \mathbb{R}^d, \text{ for } \nu \in [0, \nu^*], \tag{12.11}$$

where each $f^\nu : [T^*, \infty] \times \mathbb{R}^d \to \mathbb{R}^d$ is (at least) continuously differentiable. Let $\phi^\nu(t, t_0, x_0)$ denote the unique solution with the initial value $x(t_0) = x_0$.

Assumption 12.3 holds if, for example, the vector fields of the ODEs (12.11) satisfy a dissipative inequality such as: there exists an $R^* > 0$ such that

$$\langle f^\nu(t,x), x \rangle \leq -1 \quad \text{for all } t \geq T^*, \|x\| \geq R^*, \nu \in [0, \nu^*].$$

In this case, B is the closed and bounded ball $B_0[R^*]$ of radius R^* about the origin in $\mathbb{R}^d$.

In addition, Assumption 12.4 holds if, for example, the vector fields of the ODEs (12.11) uniform Lipschitz condition (12.6) uniformly in the parameter ν and converge continuously as the parameter $\nu \to 0$ uniformly in time on the set B, i.e., if for every $\varepsilon > 0$ and $T > 0$ there exists a $\widehat{\delta}(\varepsilon, T) > 0$ such that

$$\left\| f^\nu(t,x) - f^0(t,x) \right\| < \varepsilon, \qquad t_0 \leq t \leq t_0 + T, x \in B, \tag{12.12}$$

for $|\nu| < \widehat{\delta}(\varepsilon, T)$ and any $t_0 \geq T^*$.

Consider the solutions $\phi^0(t, \tau, x_0)$ and $\phi^\nu(t, \tau, x_0)$ of the ODEs (12.5) and (12.11), respectively, in the set B. Then

$$\left\| \phi^\nu(t, \tau, x_0) - \phi^0(t, \tau, x_0) \right\| \leq \int_\tau^t \left\| f^\nu(s, \phi^\nu(s, \tau, x_0))) - f^0(s, \phi^0(s, \tau, x_0)) \right\| ds$$

$$\leq \int_\tau^t \left\| f^\nu(s, \phi^\nu(s, \tau, x_0)) - f^\nu(s, \phi^0(s, \tau, x_0)) \right\| ds$$

$$+ \int_\tau^t \left\| f^\nu(s, \phi^0(s, \tau, x_0)) - f^0(s, \phi^0(s, \tau, x_0)) \right\| ds$$

$$\leq L_B \int_\tau^t \left\| \phi^\nu(s, \tau, x_0) - \phi^0((s, \tau, x_0) \right\| ds + \varepsilon T,$$

when $0 \leq t - \tau \leq T$. Here L_B is the common Lipschitz constant, while the continuous convergence(12.12) is used to estimate the integral in the second last line. Then Gronwall's inequality gives

$$\left\| \phi^\nu(t, \tau, x_0) - \phi^0(t, \tau, x_0) \right\| \leq \varepsilon T e^{L_B(t-\tau)} \leq \varepsilon T e^{L_B T},$$

when $0 \leq t - \tau \leq T$, i.e., the bound depends just on the length of the time interval and not the starting time.

Then the assertion of Assumption 12.4 then holds with $\delta(\varepsilon, T) = \widehat{\delta}(\varepsilon T^{-1} e^{-L_B T}, T)$.

Example 12.5. Continuing with Example 12.3, but now with a parameterised family of mappings $f_n^\nu : \mathbb{R}^d \to \mathbb{R}^d$, we note that Assumption 12.4 holds of the f_n are Lipschitz on B uniformly in $n \geq N^*$, i.e., there exists a constant L such that

$$\max_{b \in B} \left\| f_n^\nu(b) - f_n^0(b) \right\| \to 0 \quad \text{as } \nu \to 0 \text{ uniformly in } n \geq N^*.$$

12.4 Concluding remark

The nonautonomity of a dynamical system allows a much wider variety of behaviour than is possible in an autonomous system. Essentially, the vector fields of the underlying differential and difference equations can vary arbitrarily in time.

In order to obtain specific results about the behaviour of such systems, as we have seen, we need to restrict the arbitrariness of these changes in some way with some kind of uniformity assumption such as periodicity, almost periodicity or some eventual uniformity in attraction rates, etc.

Although such assumptions restrict the classes of systems under consideration, these classes are, nevertheless, still of interest in a wide range of applications.

PART 4
Random attractors

Chapter 13

Random dynamical systems

Random dynamical systems are intrinsically nonautonomous due to the changing nature of the driving noise. Continuous time random dynamical systems are generated by random ordinary differential equations (RODEs) that is, ordinary differential equations with random coefficients or stochastic processes in their coefficients, and by stochastic differential equations (SDEs), while random difference equations generate discrete time random dynamical systems. Their solutions are stochastic processes, which can be thought of as a family of time dependent paths labelled by a point ω in sample space Ω of the underlying probability space $(\Omega, \mathcal{F}, \mathbb{P})$.

A very fundamental stochastic process is the Wiener process or Brownian motion $\{W_t\}_{t \geq 0}$.

Definition 13.1. A Wiener process $\{W_t\}_{t \geq 0}$ is defined by the properties:

1) $W_t \sim N(0, t)$ for each $t \geq 0$, i.e., W_t is Gaussian distributed with

    (i) $W_0 = 0$ with probability 1;
    (ii) $\mathbb{E}(W_t) = 0$ for each $t \geq 0$;
    (iii) $\mathbb{E}(W_t^2) = t$ for each $t \geq 0$;

2) the nonoverlapping increments of W_t are independent, i.e.,

$$W_{t_2} - W_{t_1} \text{ and } W_{t_4} - W_{t_3}$$

are independent random variables for all $0 \leq t_1 < t_2 \leq t_3 < t_4$.

It follows from these properties that a Wiener process has continuous, in fact Hölder continuous sample paths, which are not differentiable. Other important stochastic processes with continuous sample paths are the fractional Brownian motion and the Ornstein-Uhlenbeck process.

A two-sided Wiener process W_t is defined for all $t \in \mathbb{R}$ and not just for $\mathbb{R}^+$. Essentially, W_t and W_{-t} for $t \in \mathbb{R}^+$ are two independent Wiener processes as in the definition above.

13.0.1   *A simple example*

RODEs are essentially nonautonomous ODEs for each path labelled by $\omega \in \Omega$ and can be solved using deterministic calculus and the standard methods and tricks for ODES.

Let $\cos W_t(\omega)$ be a specific sample path of a two-sided Wiener process W_t and consider the simple scalar RODE

$$\frac{dx}{dt} = -x + \cos W_t(\omega), \tag{13.1}$$

which obviously has no constant solution. However, the RODE (13.1) with an initial value $x_0 \in \mathbb{R}^1$ at the initial time $t_0 \in \mathbb{R}$ has the explicit solution

$$x(t, t_0, x_0, \omega) = x_0 e^{-(t-t_0)} + e^{-t} \int_{t_0}^{t} e^s \cos W_s(\omega)\, ds. \tag{13.2}$$

This is very similar to Example 5.4, but with bounded noise $\cos W_t(\omega)$ instead of the deterministic $\cos t$ as the forcing term. It has no pathwise limit as $t \to \infty$, but the difference of two solutions satisfies

$$|x(t, t_0, x_0, \omega) - y(t, t_0, y_0, \omega)| \leq |x_0 - y_0|\, e^{-(t-t_0)} \to 0 \quad \text{as } t \to \infty, \tag{13.3}$$

so all solutions converge pathwise to each other.

On the other hand, the pullback limit, i.e., for $t_0 \to -\infty$ with t held fixed, of the solution (13.2) does exist and is given by

$$\lim_{t_0 \to -\infty} x(t, \omega) = \bar{x}(t, \omega) := e^{-t} \int_{-\infty}^{t} e^s \cos W_s(\omega)\, ds. \tag{13.4}$$

Moreover, $\bar{x}(t, \omega)$ is itself a solution of the RODE (13.1), so the inequality (13.3) applies and gives

$$|x(t, t_0, x_0, \omega) - \bar{x}(t, \omega)| \leq |x_0 - \bar{x}(t_0, \omega)|\, e^{-(t-t_0)} \to 0 \quad \text{as } t \to \infty. \tag{13.5}$$

This means that the pullback limit solution (13.4) attracts all other solutions for each sample path forwards in time. It is, pathwise, globally asymptotically stable with an exponential rate of attraction. It is the counterpart of a nonautonomous equilibrium solution in a nonautonomous ODE and is sometimes called a *random equilibrium*.

Remark 13.1. This simple example suggests that random dynamical systems could be defined analogously to nonautonomous 2-parameter semi-groups with an additional parameter $\omega \in \Omega$ to label the sample paths. Pullback attractors of such random dynamical systems would then be a candidate for the definition of a random attractor. In the above example the random attractor would consist of the singleton sets $A(t, \omega) = \{\bar{x}(t, \omega)\}$ for $t \in \mathbb{R}$ and $\omega \in \Omega$.

This approach is particularly useful when the system is also driven by deterministic forcing, e.g., periodic forcing, as well as random forcing as for example in the RODE

$$\frac{dx}{dt} = -x + \cos t + \cos W_t(\omega), \tag{13.6}$$

where W_t is two-sided Wiener process. This has the explicit solution

$$x(t, t_0, x_0, \omega) = x_0 e^{-(t-t_0)} + e^{-t} \int_{t_0}^t e^s \cos s \, ds + e^{-t} \int_{t_0}^t e^s \cos W_s(\omega) \, ds \quad (13.7)$$

and the pullback limit of the solution (13.7) is

$$\lim_{t_0 \to -\infty} x(t, \omega) = \bar{x}(t, \omega) := \frac{1}{2}(\cos t + \sin t) + e^{-t} \int_{-\infty}^t e^s \cos W_s(\omega) \, ds.$$

This is a unique bounded entire solution that attracts all other solutions in both the pullback and forwards senses. It suggests that the random 2-parameter semi-group $x(t, t_0, x_0, \omega)$ generated by the RODE (13.6) has a nonautonomous random attractor which consists of the singleton sets $A(t, \omega) = \{\bar{x}(t, \omega)\}$ for $t \in \mathbb{R}$ and $\omega \in \Omega$. This approach has been used in the literature to define nonautonomous random dynamical systems and a nonautonomous random attractors.

13.1 Definition of a random dynamical system

Historically, the mathematical theory of random dynamical systems (RDS) has developed similarly to skew product flows, except that the driving system is represented by a metrical dynamical system (this terminology comes from ergodic theory) on a measure space and is measurable rather than continuous in the corresponding variable. This system represents the noise which drives the system, while a cocycle mapping represents the dynamics in the topological state space. This approach, while seemingly more complicated, has various mathematical advantages.

For example, it allows the theory to be developed without direct reference to the detailed properties of a specific driving stochastic process. Instead the driving system θ in the continuous time case is defined in terms of shift operators θ_t on the canonical sample space $\Omega := C_0(\mathbb{R}, \mathbb{R})$ of continuous functions $\omega : \mathbb{R} \to \mathbb{R}$ with $\omega(0) = 0$, i.e., with

$$\theta_t(\omega(\cdot)) := \omega(t + \cdot) - \omega(\cdot), \quad t \in \mathbb{R}. \quad (13.8)$$

The σ-algebra of Borel subsets of $C_0(\mathbb{R}, \mathbb{R})$ is taken as the σ-algebra of events $\mathcal{F}$ and $\mathbb{P}$ is the corresponding Wiener measure when noise is based on a Wiener process and similarly for other noises. Note that no other topological properties of the space $C_0(\mathbb{R}, \mathbb{R})$ are used here apart from those defining the Borel sets.

Remark 13.2. Essentially, the detailed information of the specific noise process is built into and hidden in the underlying probability space $(\Omega, \mathcal{F}, \mathbb{P})$, which allows a more general and abstract theory to be developed. The price that one pays for this generality is the difficulty in applying it to certain applications.

Let $(\Omega, \mathcal{F}, \mathbb{P})$ be a probability space, i.e., with sample space Ω, a σ-algebra $\mathcal{F}$ of admissible events (measurable subsets of Ω) and a probability measure $\mathbb{P}$ on $\mathcal{F}$, and let $\mathbb{T} = \mathbb{R}$ or $\mathbb{Z}$, depending on whether a continuous time or discrete time system is being considered.

Definition 13.2 (Random dynamical system). A *random dynamical system* (θ, φ) on $\Omega \times \mathbb{R}^d$ consists of a metric dynamical system θ on Ω, i.e., a group family of measure preserving transformations $\theta_t : \Omega \to \Omega$, $t \in \mathbb{T}$, such that

(i) $\theta_0 = \mathrm{id}_\Omega$ and $\theta_t \circ \theta_s = \theta_{t+s}$ for all $t, s \in \mathbb{T}$,
(ii) the map $(t, \omega) \mapsto \theta_t(\omega)$ is measurable and invariant with respect to $\mathbb{P}$ in the sense that $\theta_t(\mathbb{P}) = \mathbb{P}$ for all $t \in \mathbb{T}$,

and a cocycle mapping $\varphi : \mathbb{T}_0^+ \times \Omega \times \mathbb{R}^d \to \mathbb{R}^d$ such that

(a) *initial condition:* $\varphi(0, \omega, x) = x$ for all $\omega \in \Omega$ and $x \in \mathbb{R}^d$,
(b) *cocycle property:* $\varphi(s+t, \omega, x) = \varphi(s, \theta_t(\omega), \varphi(t, \omega, x))$ for all $s, t \in \mathbb{R}^+$, $\omega \in \Omega$ and $x \in \mathbb{R}^d$,
(c) *measurability:* $(t, \omega, x) \mapsto \varphi(t, \omega, x)$ is measurable,
(d) *continuity:* $x \mapsto \varphi(t, \omega, x)$ is continuous for all $(t, \omega) \in \mathbb{R} \times \Omega$.

The notation $\theta_t(\mathbb{T}) = \mathbb{T}$ for the measure preserving property of θ_t with respect to $\mathbb{T}$ is just a compact way of writing
$$\mathbb{T}(\theta_t(A)) = \mathbb{T}(A), \quad t \in \mathbb{R}, \ A \in \mathcal{F}.$$
A systematic treatment of the theory of random dynamical system is given in [Arnold (1998)], where technical details can be found.

Example 13.1. Using the canonical sample space representation (13.8) of the noise, the RODE (13.1) takes the form
$$\frac{dx}{dt} = -x + \cos(\theta_t(\omega)),$$
and its cocycle mapping for an initial value $x_0 \in \mathbb{R}^1$ and noise sample path $\omega \in \Omega$ is given by
$$\varphi(t, x_0, \omega) = x_0 e^{-t} + e^{-t} \int_0^t e^s \cos(\theta_s(\omega)) \, ds,$$
that is, $\varphi(t, x_0, \omega) = x(t, 0, x_0, \omega)$. Here t is the time that has elapsed since starting with the initial noise state ω and initial value $x_0 \in \mathbb{R}^1$.

The solution $x(t, t_0, x_0, \omega)$ of the RODE (13.1) given by (13.2) corresponds to $\varphi(t, x_0, \theta_{t_0}(\omega))$ with the initial noise state ω adjusted to $\theta_{t_0}(\omega)$ corresponding to the starting time t_0.

Remark 13.3. A random dynamical system (θ, φ) on $\Omega \times \mathbb{R}^d$ is not a skew product flow or an autonomous semi-dynamical system on $\Omega \times \mathbb{R}^d$, since Ω (in the context considered here) is not a topological space. Nevertheless, skew product flows and random dynamical systems have many common properties, and concepts and results for one can be used with appropriate modifications for the other. The most significant modification concerns measurability and the use of families of random sets.

13.2   Random attractors

The example in the previous section suggests that a random attractor can be defined similarly to pullback attractors or skew product flows in Chapter 9. This requires the following definitions.

Definition 13.3. A family $\mathcal{D} = \{D(\omega), \omega \in \Omega\}$ of nonempty subsets of $\mathbb{R}^d$ is called a *random set* if the mapping $\omega \mapsto \text{dist}_{\mathbb{R}^d}(x, D(\omega))$ is $\mathcal{F}$-measurable for all $x \in \mathbb{R}^d$.

A random set $\mathcal{D} = \{D(\omega), \omega \in \Omega\}$ is called a *random closed set* if $D(\omega)$ is closed for each $\omega \in \Omega$ and a *random compact set* if $D(\omega)$ is compact for each $\omega \in \Omega$.

Definition 13.4. A random set $\mathcal{D} = \{D(\omega), \omega \in \Omega\}$ is said to be *tempered* if

$$D(\omega) \subset \{x \in \mathbb{R}^d \ : \ \|x\| \leq r(\omega)\}, \quad \omega \in \Omega,$$

where the random variable $r(\omega) > 0$ is tempered, i.e.,

$$\sup_{t \in \mathbb{R}} r(\theta_t(\omega)) e^{-\gamma|t|} < \infty, \quad \gamma > 0, \ \omega \in \Omega.$$

The collection of all tempered random sets in $\mathbb{R}^d$ will be denoted by $\mathfrak{D}$.

A random attractor of a random dynamical system is essentially a pullback attractor which is a random set.

Definition 13.5 (Random attractor). A random closed set $\mathcal{A} = \{A(\omega), \omega \in \Omega\}$ from $\mathfrak{D}$ is called a *random attractor* of a random dynamical system (θ, φ) on $\Omega \times \mathbb{R}^d$ in $\mathfrak{D}$ if $\mathcal{A}$ is a φ-invariant set, i.e.,

$$\varphi(t, \omega, A(\omega)) = A(\theta_t(\omega)), \quad t \geq 0, \omega \in \Omega,$$

and pathwise pullback attracting in $\mathfrak{D}$, i.e.,

$$\lim_{t \to \infty} \text{dist}_{\mathbb{R}^d}(\varphi(t, \theta_{-t}(\omega), D(\theta_{-t}(\omega))), A(\omega)) = 0, \quad \omega \in \Omega, \ \mathcal{D} \in \mathfrak{D}.$$

The existence of a random attractor follows from that of a pullback absorbing tempered random set, i.e., a tempered random set $\mathcal{B} = \{B(\omega), \omega \in \Omega\}$ with compact component sets for which there exists a $T_{\mathcal{D},\omega} \geq 0$ such that

$$\varphi(t, \theta_{-t}(\omega), D(\theta_{-t}(\omega)) \subset B(\omega), \quad t \geq T_{\mathcal{D},\omega}, \tag{13.9}$$

for every tempered random set $\mathcal{D} = \{D(\omega), \omega \in \Omega\}$.

Theorem 13.1 (Existence of random attractors). *Let (θ, φ) be a random dynamical system on $\Omega \times \mathbb{R}^d$ such that $\varphi(t, \omega, \cdot) : \mathbb{R}^d \to \mathbb{R}^d$ is a continuous for each fixed $t > 0$ and $\omega \in \Omega$. If there exists a tempered random set $\mathcal{B} = \{B(\omega), \omega \in \Omega\}$ with compact component sets, then the random dynamical system (θ, φ) has a random pullback attractor $\mathcal{A} = \{A(\omega), \omega \in \Omega\}$ with compact component sets defined for each $\omega \in \Omega$ by*

$$A(\omega) = \bigcap_{s>0} \overline{\bigcup_{t \geq s} \varphi(t, \theta_{-t}(\omega), B(\theta_{-t}(\omega)))}. \tag{13.10}$$

Proof. The proof of Theorem 13.1 is essentially the same as for its counterpart Theorem 9.1 for deterministic skew product flows, see the proof of Theorem 3.20 in [Kloeden and Rasmussen (2011)]. The full proof in the random case is given in [Crauel and Flandoli (1994)].

The only new feature is that of measurability, i.e., to show that $\mathcal{A} = \{A(\omega), \omega \in \Omega\}$ is a random set. This follows from the fact that the set-valued mappings $\omega \mapsto \varphi(t, \theta_{-t}(\omega), B(\theta_{-t}(\omega)))$ are measurable for each $t \in \mathbb{R}_0^+$ and that the union and intersection in (13.10) can be taken over a countable number of time instants (in the continuous time case).   □

Remark 13.4. There is no analogous theorem for the existence of a random forward attractor, although when θ is ergodic, then

$$\mathbb{P}\left(\omega \in \Omega \ : \ \mathrm{dist}_{\mathbb{R}^d}\left(\varphi(t, \theta_{-t}(\omega), B(\theta_{-t}(\omega)), A(\omega)\right) \geq \varepsilon\right)$$

$$= \mathbb{P}\left(\omega \in \Omega \ : \ \mathrm{dist}_{\mathbb{R}^d}\left(\varphi(t, \omega, B(\omega), A(\theta_t(\omega))\right) \geq \varepsilon\right)$$

for each $\varepsilon > 0$, so a (pathwise convergent, hence convergent in probability) random pullback attractor also converges in the forwards direction, but in the weaker sense of convergence in probability. This is due to the possibility of large deviations of individual sample paths.

Example 13.2. If the random attractor consists of singleton sets, i.e., $A(\omega) = \{X^*(\omega)\}$ for some random variable X^* with values in $\mathbb{R}^d$, then $\bar{X}_t(\omega) := X^*(\theta_t(\omega))$ is a stationary stochastic process on $\mathbb{R}^d$. In Example 13.1 above, X^* is given by the pullback limit, i.e.,

$$X^*(\omega) := \int_{-\infty}^0 e^s \cos W_s(\omega)\, ds,$$

which is essentially the limit (13.4) at time $t = 0$. In this case the random attractor is also forward attracting in the stronger pathwise sense.

The example above is a strictly contracting system. There is a random counterpart of Theorem 7.2, which gives the existence of a random attractor consisting of singleton subsets for RDS with the uniformly strictly contracting property, see [Caraballo, Kloeden and Schmalfuß (2004)].

13.3   Examples of random attractors

13.3.1   *Random attractors for RODES*

To apply the theory of random dynamical systems to RODEs, the RODEs need to be rewritten in the canonical form

$$\frac{dx}{dt} = f(x, \theta_t(\omega)), \qquad x \in \mathbb{R}^d, \tag{13.11}$$

with a the noise represented by a measure-preserving dynamical system $\theta = \{\theta_t\}_{t\in\mathbb{R}}$ acting on a probability space $(\Omega, \mathcal{F}, \mathbb{P})$ rather than by a specific noise process as in (13.1).

This was done in Example 13.1 above, which has a simple random attractor consisting of singleton subsets. For a less trivial random attractor consider the Bernoulli ODE (8.3), i.e.,

$$\frac{dx}{dt} = ax - b(t)x^3,$$

see Subsection 8.3, with the coefficient function $b(t)$ replaced by a random term $b(\theta_t(\omega))$ to give the RODE

$$\frac{dx}{dt} = ax - b(\theta_t(\omega))x^3, \tag{13.12}$$

where $a > 0$ and $b : \Omega \to \mathbb{R}$ is continuous and satisfies $b(\omega) \in [b_0, b_1]$ for all $\omega \in \Omega$, where $0 < b_0 < b_1 < \infty$. An example is $b(\theta_t(\omega)) = 1 + \cos^2(\theta_t(\omega))$ with the noise paths coming from a Wiener process.

It can be shown in the same way as in Subsection 8.3 with $b(t)$ representing $b(\theta_t(\omega))$ that the random dynamical system generated by the RODE (13.12) has random equilbria solutions $\pm\bar{\phi}_a(\omega)$, where

$$\bar{\phi}_a(\omega) = \frac{1}{\sqrt{2\int_{-\infty}^0 b(\theta_s(\omega))e^{-2as}\,ds}}, \quad \omega \in \Omega,$$

is obtained by taking the pathwise limit, and that the random attractor $\mathcal{A} = \{A(\omega), \omega \in \Omega\}$ here has the component interval sets

$$A_a(\omega) = \left[-\bar{\phi}_a(\omega), \bar{\phi}_a(\omega)\right], \quad \omega \in \Omega. \tag{13.13}$$

13.3.2 Random attractors for stochastic differential equations

Stochastic differential equations are not differential equations at all since the sample paths of their solutions, like those of the driving Wiener process, are not differentiable. They are in fact integral equations and require the stochastic calculus for their treatment rather than the familiar deterministic calculus.

The general form of a scalar stochastic differential equation is

$$dX_t = f(t, X_t)dt + g(t, X_t)dW_t,$$

where f is called the drift coefficient and g the diffusion coefficient. In addition, W_t is a scalar Wiener process, which is assumed here to be two-sided, i.e., defined for all $t \in \mathbb{R}$. This is only a symbolic representation of the stochastic integral equation

$$X_t = X_{t_0} + \int_{t_0}^t f(s, X_s)\,ds + \int_{t_0}^t g(s, X_s)\,dW_s,$$

where the second integral is an Itô stochastic integral.

The cocycle mapping is defined by $\varphi(t, \omega, x_0) := X_t^{x_0}(\omega)$, where $X_t^{x_0}$ is the solution of the SDE with deterministic initial value x_0 at time $t = 0$, and θ_t is the shift operator on the canonical sample space $(\Omega, \mathcal{F}, \mathbb{P})$ defined in terms of the sample paths of the Wiener process.

It is not difficult to show that the mapping φ so defined satisfies the cocycle property for all ω except those in a null event $N_{s,t} \in \mathcal{F}$, i.e., with $\mathbb{P}(N_{s,t}) = 0$, for given $s, t \in \mathbb{R}_0^+$. This gives the *crude* cocycle property. However, Definition 13.2 requires the *perfect* cocycle property, for which the relationship holds for all ω except those in a null event N which is independent of s, $t \in \mathbb{R}_0^+$. The original sample space Ω is then replaced by $\Omega^* := \Omega \setminus N$. It is much harder to show that the perfect cocycle property holds, see [Arnold (1998)].

13.3.2.1 *The Ornstein–Uhlenbeck process*

The Ornstein–Uhlenbeck process provides a simple example of a random attractor, which can be computed explicitly. It is the unique stochastic stationary solution a linear Itô stochastic differential equation (SDE) with additive noise and plays an important role in investigating random attractors in nonlinear SDE.

Consider the SDE with linear drift and additive noise,

$$dX_t = -X_t dt + \alpha dW_t, \tag{13.14}$$

where α is a positive constant, which has the explicit solution

$$X_t = X_{t_0} e^{-(t-t_0)} + \alpha e^{-t} \int_{t_0}^{t} e^s \, dW_s.$$

Let W_t is a two-sided Wiener process, i.e., defined for all $t \in \mathbb{R}$. Taking the pathwise pullback limit as $t_0 \to -\infty$ gives the Ornstein–Uhlenbeck process

$$O_t = \alpha e^{-t} \int_{-\infty}^{t} e^s \, dW_s, \tag{13.15}$$

which is the unique stochastic stationary solution of the SDE (13.14). In particular, $O_t(\omega) = O^*(\theta_t(\omega))$, where

$$O^*(\omega) := \int_{-\infty}^{0} e^s W_s(\omega) \, ds.$$

The SDE (13.14) is strictly contracting and all other solutions converge pathwise to the Ornstein–Uhlenbeck solution forwards in time. Subtracting any two solutions X_t and Y_t of (13.14) gives

$$\frac{d}{dt}(X_t - Y_t) = -(X_t - Y_t),$$

so, pathwise,

$$|X_t - Y_t| = |x_0 - y_0| e^{-t} \to 0 \quad \text{as } t \to \infty.$$

Its random attractor $\mathcal{A} = \{A(\omega), \omega \in \Omega\}$ thus has singleton sets $A(\omega) = \{O^*(\omega)\}$.

13.3.2.2 *A nonlinear SDE with additive noise*

Consider a nonlinear scalar SDE with additive noise

$$dX_t = f(X_t)dt + \alpha dW_t, \tag{13.16}$$

where α is a positive constant. Suppose that the drift coefficient $f : \mathbb{R} \to \mathbb{R}$ is continuously differentiable and satisfies the one-sided dissipative Lipschitz condition

$$\langle x_1 - x_2, f(x_1) - f(x_2) \rangle \leq -L|x_1 - x_2|^2, \quad x_1, x_2 \in \mathbb{R}, \tag{13.17}$$

and also satisfies

Assumption 13.1. Integrability condition: There exists $\lambda_0 > 0$ such that

$$\int_{-\infty}^{t} e^{\lambda s} |f(u(s))|^2 \, ds < \infty \tag{13.18}$$

holds for every $\lambda \in (0, \lambda_0]$ and any continuous function $u : \mathbb{R} \to \mathbb{R}^d$ with sub-exponential growth. (It can be assumed without loss of generality that $L \leq m_0$).

The SDE (13.16) is really the integral equation

$$X_t = X_0 + \int_0^t f(X_s) \, ds + \alpha W_t.$$

In order to use the one-sided dissipative Lipschitz condition (13.17) we need to transform it into a RODE. Consider the difference $X_t - O_t$, where O_t is the Ornstein–Uhlenbeck stationary process (13.15) satisfying the linear stochastic differential equation (13.14). This difference is pathwise continuous since the paths are continuous and satisfy the integral equation

$$X_t - O_t = X_0 - O_0 + \int_0^t (f(X_s) + O_s) \, ds,$$

so by the Fundamental Theorem of Calculus the difference is actually pathwise differentiable.

This integral equation is thus equivalent to the pathwise RODE

$$\frac{dx}{dt} = f(x + O_t) + O_t,$$

where $x(t, \omega) := X_t(\omega) - O_t(\omega)$, or more simply,

$$\frac{d}{dt}(X_t - O_t) = f(X_t) + O_t$$

for each $\omega \in \Omega$.

Now take the inner product with $X_t - O_t$ and apply the one-sided Lipschitz condition to this RODE to obtain

$$\frac{d}{dt}|X_t - O_t|^2 = 2\langle X_t - O_t, f(X_t) - f(O_t) \rangle + 2\langle X_t - O_t, f(O_t) + O_t \rangle$$

$$\leq -2L|X_t - O_t|^2 + L|X_t - O_t|^2 + \frac{4}{L}|f(O_t) + O_t|^2.$$

Hence,

$$|X_t - O_t|^2 \leq |X_{t_0} - O_{t_0}|^2 e^{-L(t-t_0)} + \frac{4e^{-Lt}}{L} \int_{t_0}^t e^{Ls} |f(O_s) + O_s|^2 \, ds \, .$$

Pathwise pullback convergence (i.e., as $t_0 \to -\infty$) then gives

$$|X_t - O_t|^2 \leq R_X^2(\theta_t(\omega)) := 1 + \frac{4e^{-Lt}}{L} \int_{-\infty}^t e^{Ls} |f(O(\theta_s(\omega))) + O(\theta_s(\omega))|^2 \, ds$$

for all $t \geq T_\mathcal{D}$ and tempered families $\mathcal{D} = (D_\omega)_{\omega \in \Omega}$ of the initial conditions. The integrability condition (13.18) and properties of the Ornstein–Uhlenbeck process ensure that $R_X^2(\theta_t(\omega))$ is finite.

Thus,

$$|X_t(\omega) - O_t(\omega)| \leq R_X(\theta_t(\omega)), \quad t \geq T_\mathcal{D} \, ,$$

or equivalently

$$|X_t(\omega)| \leq |O_t(\omega)| + R_X(\theta_t(\omega)), \quad t \geq T_\mathcal{D} \, .$$

The family of compact balls $B(\omega)$ centered on $O_0(\omega)$ with radius $R_X(\omega)$ is a tempered pullback absorbing family of compact subsets. This means that this system has a random attractor $\mathcal{A} = \{A(\omega), \omega \in \Omega\}$ by Theorem 13.1.

The difference of any two solutions satisfies the differential inequality

$$\frac{d}{dt} |X_t^1 - X_t^2|^2 \leq -2L |X_t^1 - X_t^2|^2 \, ,$$

so the system is strictly contracting. This means all solutions converge pathwise to each other forwards in time and that the random attractor subsets are singleton sets $A(\omega) = \{X^*(\omega)\}$, i.e., the random attractor is formed by a stationary stochastic process $\bar{X}_t(\omega) = X^*(\theta_t(\omega))$.

Chapter 14

Mean-square random dynamical systems

Mean-square properties are of traditional interest in the investigation of stochastic systems in engineering and physics. This is quite natural since the Itô stochastic calculus is a mean-square calculus. At first sight, it is thus somewhat surprising that the classical theory of random dynamical systems is a pathwise theory, although this can be justified since stochastic differential equations can be transformed into pathwise random ordinary differential equations as in the example at the end of the last chapter.

Such transformations, however, need not apply to mean-field stochastic differential equations of the form

$$dX_t = f(t, X_t, \mathbb{E}X_t, \mathbb{E}X_t^2)\,dt + g(t, X_t, \mathbb{E}X_t, \mathbb{E}X_t^2)\,dW_t \qquad (14.1)$$

which include expectations and higher moments of the solution in their coefficient functions.

Let $\{W_t\}_{t \in \mathbb{R}}$ be a two-sided scalar Wiener process defined on a probability space $(\Omega, \mathcal{F}, \mathbb{P})$ and let $\{\mathcal{F}_t\}_{t \in \mathbb{R}}$ be the natural filtration generated by $\{W_t\}_{t \in \mathbb{R}}$, i.e., an increasing and right-continuous family of sub-σ-algebras of $\mathcal{F}$, which contain all $\mathbb{P}$-null sets. Essentially, $\mathcal{F}_t$ represents the information about the randomness at time $t \in \mathbb{R}$. Define the spaces

$$\mathfrak{X} := L^2(\Omega, \mathcal{F}; \mathbb{R}^d), \quad \mathfrak{X}_t := L^2(\Omega, \mathcal{F}_t; \mathbb{R}^d) \quad \text{for } t \in \mathbb{R}$$

with the norm $\|X\|_{\mathrm{ms}} := \sqrt{\mathbb{E}\|X\|^2}$, where $\|\cdot\|$ is the Euclidean norm on $\mathbb{R}^d$ and note that

$$\mathfrak{X}_t \subset \mathfrak{X}_s \subset \mathfrak{X}, \quad -\infty < t < s < \infty.$$

Given any initial condition $X_s \in \mathfrak{X}_s$, $s \in \mathbb{R}$, a solution of (14.1) is a stochastic process $\{X_t\}_{t \geq s}$ with $X_t \in \mathfrak{X}_t$ for $t \geq s$, satisfying the stochastic integral equation

$$X_t = X_s + \int_s^t f(u, X_u, \mathbb{E}X_u, \mathbb{E}X_u^2)\,du + \int_s^t g(u, X_u, \mathbb{E}X_u, \mathbb{E}X_u^2)\,dW_u.$$

Under appropriate assumptions on the coefficient functions, e.g., locally Lipschitz with a growth bound or a dissipativity property, the mean-field SDE (14.1) has a unique solution and generates a mean-square random dynamical system ϕ on the underlying phase space $\mathbb{R}^d$ with a probability set-up $(\Omega, \mathcal{F}, \{\mathcal{F}_t\}_{t \in \mathbb{R}}, \mathbb{P})$, defined by

$$\phi(t, s, X_s) = X_t \quad \text{for } -\infty < t < s < \infty, \ X_s \in \mathfrak{X}_s.$$

14.1   Mean-square random dynamical systems

Mean-square random dynamical systems were defined in [Kloeden and Lorenz (2012)] as deterministic two-parameter semi-groups acting on a state space of mean-square random variables.

Consider the time set $\mathbb{R}$, and define $\mathbb{R}^2_{\geq} := \{(t,s) \in \mathbb{R}^2 : t \geq s\}$.

Definition 14.1. A *mean-square random dynamical system* (MS-RDS for short) ϕ on the underlying phase space $\mathbb{R}^d$ with the filtered probability space $(\Omega, \mathcal{F}, \{\mathcal{F}_t\}_{t \in \mathbb{R}}, \mathbb{P})$ is a family of mappings

$$\phi(t, s, \cdot) : \mathfrak{X}_s \to \mathfrak{X}_t, \quad (t,s) \in \mathbb{R}^2_{\geq},$$

which satisfies:

(1) *Initial condition:* $\phi = X_s$ for all $X_s \in \mathfrak{X}_s$ and $s \in \mathbb{R}$.
(2) *Two-parameter semigroup property:* For all $X \in \mathfrak{X}_u$ and all $(t,s), (s,u) \in \mathbb{R}^2_{\geq}$

$$\phi(t, u, X) = \phi(t, s, \phi(s, u, X)).$$

(3) *Continuity:* ϕ is continuous in its variables.

Mean-square random dynamical systems are essentially deterministic with the stochasticity built into or hidden in the time-dependent state spaces.

14.2   Mean-square random attractors

The mean-square random attractor of a mean-square random dynamical system is defined here as the pullback attractor of a deterministic nonautonomous dynamical system as in Chapter 6.

Definition 14.2. A family $\mathcal{A} = \{A(t), \, t \in \mathbb{R}\}$ of nonempty compact subsets of $\mathfrak{X}$ with $A(t) \subset \mathfrak{X}_t$ for each $t \in \mathbb{R}$ is called a *mean-square attractor* of a MS-RDS ϕ if it pullback attracts all uniformly bounded families $\mathcal{D} = \{D(t), \, t \in \mathbb{R}\}$ of subsets of $\{\mathfrak{X}_t\}_{t \in \mathbb{R}}$, i.e.,

$$\lim_{s \to -\infty} \text{dist}_{\mathfrak{X}_t} (\phi(t, s, D(s)), A(t)) = 0.$$

Uniformly bounded here means that there is an $R > 0$ such that $\|X\|_{\text{ms}} \leq R$ for all $X \in D(t)$ and $t \in \mathbb{R}$.

The existence of pullback attractors follows from that of an absorbing family.

Definition 14.3. A uniformly bounded family $\mathcal{B} = \{B(t), \, t \in \mathbb{R}\}$ of nonempty closed and bounded subsets of $\{\mathfrak{X}_t\}_{t \in \mathbb{R}}$ is called a *pullback absorbing family* for a MS-RDS ϕ if for each $t \in \mathbb{R}$ and every uniformly bounded family $\mathcal{D} = \{D(t), \, t \in \mathbb{R}\}$ of nonempty subsets of $\{\mathfrak{X}_t\}_{t \in \mathbb{R}}$, there exists some $T = T(t, \mathcal{D}) \in \mathbb{R}^+$ such that

$$\phi(t, s, D(s)) \subseteq B(t) \quad \text{for } s \in \mathbb{R} \text{ with } s \leq t - T.$$

Since closed and bounded sets in the infinite dimensional space $\mathfrak{X}$ are not necessarily compact, the mappings defining the MS-RDS must be assumed to have some kind of compactness property.

Theorem 14.1. *Suppose that a MS-RDS ϕ has a positively invariant pullback absorbing uniformly bounded family $\mathcal{B} = \{B(t), t \in \mathbb{R}\}$ of nonempty closed and bounded subsets of $\{\mathfrak{X}_t\}_{t\in\mathbb{R}}$ and that the mappings $\phi(t, s, \cdot) : \mathfrak{X}_s \to \mathfrak{X}_t$ are pullback compact (respectively, eventually or asymptotically compact) for all $(t, s) \in \mathbb{R}_{\geq}^2$. Then, ϕ has a unique mean-square random attractor $\mathcal{A} = \{A(t), t \in \mathbb{R}\}$ with component sets determined by*

$$A(t) = \bigcap_{s \leq t} \phi(t, s, B(s)), \quad t \in \mathbb{R}.$$

The main difficulty in applying this theorem is the lack of useful characterisations of compact sets of such spaces $\mathfrak{X}$ of mean-square random variables. In the example below total boundedness, hence pre-compactness, will be shown directly.

14.3 Example of a mean-square attractor

Consider the nonlinear mean-field SDE

$$dX_t = \left(\alpha X_t + \mathbb{E}X_t - X_t \mathbb{E}X_t^2\right) dt + X_t \, dW_t \tag{14.2}$$

with a real-valued parameter α.

The first and second moment equations of the mean-field SDE (14.2) are given by

$$\frac{d}{dt}\mathbb{E}X_t = (\alpha + 1)\mathbb{E}X_t - \mathbb{E}X_t\mathbb{E}X_t^2, \tag{14.3}$$

$$\frac{d}{dt}\mathbb{E}X_t^2 = (2\alpha + 1)\mathbb{E}X_t^2 + 2(\mathbb{E}X_t)^2 - 2(\mathbb{E}X_t^2)^2, \tag{14.4}$$

where Itô's formula (i.e., the chain rule in stochastic calculus) with $y = x^2$ was used to derive (14.4). These can be rewritten as the system of ODEs

$$\frac{dx}{dt} = x(\alpha + 1 - y)$$

$$\frac{dy}{dt} = (2\alpha + 1)y + 2x^2 - 2y^2$$

under the restriction that $x^2 \leq y$.

This system of ODEs has a steady state solution $\bar{x} = \bar{y} = 0$ for all α, which corresponds to the zero solution $X_t \equiv 0$ of the mean-field SDE (14.2). There also exist valid (i.e., with $y \geq 0$) steady state solutions $\bar{x} = \pm\sqrt{(\alpha + 1)/2}$, $\bar{y} = \alpha + 1$ for $\alpha > -1$ and $\bar{x} = 0$, $\bar{y} = \alpha + \frac{1}{2}$ for $\alpha \geq -\frac{1}{2}$.

It needs to be shown if there are solutions of the SDE (14.2) with moments corresponding to these steady state solutions.

Theorem 14.2. *The MS-RDS ϕ generated by (14.2) has a uniformly bounded positively invariant pullback absorbing family.*

Proof. Let α be arbitrary and define

$$B(t) = \left\{ X \in \mathfrak{X}_t : \|X\|_{\mathrm{ms}} \le \sqrt{|\alpha| + 2} \right\} \quad \text{for } t \in \mathbb{R}.$$

Using

$$(2\alpha + 1)\mathbb{E}X_t^2 + 2(\mathbb{E}X_t)^2 - 2(\mathbb{E}X_t^2)^2 \le (2\alpha + 3)\mathbb{E}X_t^2 - 2(\mathbb{E}X_t^2)^2,$$

which holds since $(\mathbb{E}X_t)^2 \le \mathbb{E}X_t^2$, the second moment equation (14.4) gives the differential inequality

$$\frac{d}{dt}\mathbb{E}X_t^2 \le (2\alpha + 3)\mathbb{E}X_t^2 - 2(\mathbb{E}X_t^2)^2. \tag{14.5}$$

Let $\mathcal{D} = \{D(t), t \in \mathbb{R}\}$ be a uniformly bounded family of nonempty subsets of $\{\mathfrak{X}_t\}_{t\in\mathbb{R}}$, i.e., $D(t) \subset \mathfrak{X}_t$ and there exists $R > 0$ such that $\|X\|_{\mathrm{ms}} \le R$ for all $X \in D(t)$. Specifically, it will be shown that $\phi(t, s, D(s)) \subset B(t)$ for $t - s \ge T$, where T is defined by

$$T := \log\left(\frac{R^2}{|\alpha| + 2}\right). \tag{14.6}$$

Pick $X_s \in D(s)$ arbitrarily and $(t, s) \in \mathbb{R}^2_{\ge}$ with $t - s \ge T$. Motivated by the differential inequality (14.5), consider the scalar ODE

$$\dot{y} = (2\alpha + 3)y - 2y^2, \tag{14.7}$$

with the initial value $y(s) \le R^2$. A direct computation yields

$$y(t) = y(s)\exp\left(\int_s^t (2\alpha + 3) - 2y(u)\,du\right) \le R^2 \exp\left(\int_s^t (2\alpha + 3) - 2y(u)\,du\right).$$

From the definition of T in (14.6), it follows that $\min_{s\le u\le t} y(u) \le |\alpha| + 2$.

Furthermore, $y = 0$ and $y = \alpha + \frac{3}{2}$ are stationary points of the ODE (14.7). For this reason, $\min_{s\le u\le t} y(u) \le |\alpha| + 2$ implies that $y(t) \le |\alpha| + 2$. Then from (14.5), it follows that $y(t) \ge \|\phi(t, s, X_s)\|_{\mathrm{ms}}^2$. This means that

$$\|\phi(t, s, X_s)\|_{\mathrm{ms}}^2 \le y(t) \le |\alpha| + 2, \tag{14.8}$$

i.e., $\phi(t, s, X_s) \in B_t$ for $t - s \ge T$.

Hence, $\mathcal{B} = \{B(t), t \in \mathbb{R}\}$ is a pullback absorbing family for the MS-RDS ϕ. This family is clearly uniformly bounded and positively invariant for the MS-RDS ϕ. $\qquad\square$

Theorem 14.3. *The MS-RDS ϕ generated by (14.2) has a mean-square attractor $\mathcal{A} = \{A(t), t \in \mathbb{R}\}$ with component sets $A(t) = \{0\}$ when $\alpha < -1$.*

Proof. Let $\alpha < -1$ be arbitrary. Let $\mathcal{D} = \{D(t), t \in \mathbb{R}\}$ be a uniformly bounded family of nonempty subsets of $\{\mathfrak{X}_t\}_{t\in\mathbb{R}}$ with $\|X\|_{\mathrm{ms}} \le R$ for all $X \in D(t)$ where $R > 0$.

Let $(X_s)_{s\in\mathbb{R}}$ be an arbitrary sequence with $X_s \in D(s)$. The moment equations (14.3)–(14.4) can be written as

$$\frac{d}{dt}\mathbb{E}[\phi(t,s,X_s)] = \mathbb{E}[\phi(t,s,X_s)]\left(\alpha + 1 - \mathbb{E}[\phi(t,s,X_s)]^2\right)$$

$$\frac{d}{dt}\mathbb{E}[\phi(t,s,X_s)]^2 = (2\alpha+1)\mathbb{E}[\phi(t,s,X_s)]^2 + 2(\mathbb{E}[\phi(t,s)X_s])^2 - 2(\mathbb{E}[\phi(t,s)X_s]^2)^2.$$

Then

$$\mathbb{E}[\phi(t,s,X_s)] = \mathbb{E}X_s \exp\left(\int_s^t (\alpha + 1 - \mathbb{E}X_u^2)\,du\right),$$

which implies that

$$(\mathbb{E}[\phi(t,s,X_s)])^2 \le e^{(\alpha+1)(t-s)}R^2.$$

Moreover, by the variation of constants formula,

$$\mathbb{E}[\phi(t,s,X_s)]^2$$
$$= e^{(2\alpha+1)(t-s)}\mathbb{E}X_s^2 + 2\int_s^t e^{(2\alpha+1)(t-u)}\left(\mathbb{E}[\phi(u,s,X_s)]^2 - (\mathbb{E}[\phi(u,s,X_s)]^2)^2\right)du$$
$$\le e^{(2\alpha+1)(t-s)}R^2 + 2R^2\int_s^t e^{(2\alpha+1)(t-u)}e^{(\alpha+1)(u-s)}\,du$$
$$\le e^{(2\alpha+1)(t-s)}R^2 + 2e^{(\alpha+1)(t-s)}(t-s)R^2,$$

which implies that $\lim_{s\to-\infty}\|\phi(t,s,X_s)\|_{ms} = 0$.

Thus the family of sets $A(t) = \{0\}$ is the mean-square attractor of (14.2) in this case. □

When α becomes larger than -1, the mean-square attractor bifurcates to include nontrivial solutions.

Theorem 14.4. *The MS-RDS ϕ generated by (14.2) has a nontrivial mean-square attractor when $-1 < \alpha < -\frac{1}{2}$.*

Proof. In order to apply Theorem 14.1, we need to show that ϕ is pullback asymptotically compact, i.e., given a uniformly bounded family $\mathcal{D} = \{D(t), t \in \mathbb{R}\}$ of nonempty subsets of $\mathfrak{X}_t$ and sequences $\{t_k\}_{k\in\mathbb{N}}$ in $(-\infty,t)$ with $t_k \to -\infty$ as $k \to \infty$ and $\{X_k\}_{k\in\mathbb{N}}$ with $X_k \in D(t_k) \subset \mathfrak{X}_{t_k}$ for each $k \in \mathbb{N}$, then the subset $\{\phi(t,t_k,X_k)\}_{k\in\mathbb{N}} \subset \mathfrak{X}_t$ is relatively compact.

For this purpose, let $\epsilon > 0$ be arbitrary. We will construct a finite cover of $\{\phi(t,t_k,X_k)\}_{k\in\mathbb{N}}$ with diameter less than ϵ. Choose and fix $s \in \mathbb{R}$ with $s < t$ such that

$$4e^{(2\alpha+1)(t-s)}(|\alpha| + 2)^2 < \frac{\epsilon^2}{2}, \tag{14.9}$$

and define $Y_s^k := \phi(s,t_k,X_k)$. Using (14.8) it can be assumed without loss of generality that

$$\mathbb{E}(Y_s^k)^2 \le |\alpha| + 2 \qquad \text{for } k \in \mathbb{N}. \tag{14.10}$$

For $u \in [s,t]$, let $Y_u^k := \phi(u, s, Y_s^k)$ and consider a family of functions $f^k : [s,t] \to \mathbb{R}$ defined by $f^k(u) := \mathbb{E}Y_u^k$. By the Itô formula,

$$Y_t^k = e^{\alpha(t-s)} Y_s^k + \int_s^t e^{\alpha(t-u)} \left(\mathbb{E}Y_u^k - Y_u^k \mathbb{E}(Y_u^k)^2 \right) du + \int_s^t e^{\alpha(t-u)} Y_u^k \, dW_u,$$

from which it can be shown that $\{f^k\}_{k \in \mathbb{N}}$ is a uniformly equicontinuous sequence of functions. Hence by the Arzelà–Ascoli theorem, for $\delta := \epsilon \sqrt{\frac{\alpha}{2} + \frac{1}{4}}$, there exists a finite index set $J(\delta) \subset \mathbb{N}$ such that for any $k \in \mathbb{N}$ there exists $n_k \in J(\delta)$ for which

$$\sup_{u \in [s,t]} |\mathbb{E}Y_u^k - \mathbb{E}Y_u^{n_k}| < \delta. \tag{14.11}$$

To conclude the proof of asymptotic compactness, it is sufficient to show the inequality $\|Y_t^k - Y_t^{n_k}\|_{\mathrm{ms}} \leq \epsilon$. Indeed, for $u \in [s,t]$, a direct computation gives

$$\frac{d}{dt} \mathbb{E}(Y_u^k - Y_u^{n_k})^2 = (1 + 2\alpha)\mathbb{E}(Y_u^k - Y_u^{n_k})^2 + 2(\mathbb{E}Y_u^k - \mathbb{E}Y_u^{n_k})^2 - 2(\mathbb{E}(Y_u^k)^2)^2$$
$$- 2(\mathbb{E}(Y_u^{n_k})^2)^2 + 2\mathbb{E}Y_u^k Y_u^{n_k} (\mathbb{E}(Y_u^k)^2 + \mathbb{E}(Y_u^{n_k})^2).$$

Note that

$$2(\mathbb{E}(Y_u^k)^2)^2 + 2(\mathbb{E}(Y_u^{n_k})^2)^2 - 2\mathbb{E}Y_u^k Y_u^{n_k} (\mathbb{E}(Y_u^k)^2 + \mathbb{E}(Y_u^{n_k})^2) \geq 0.$$

Hence, by the variation of constants formula,

$$\|Y_t^k - Y_t^{n_k}\|_{\mathrm{ms}}^2 \leq e^{(2\alpha+1)(t-s)} \|Y_s^k - Y_s^{n_k}\|_{\mathrm{ms}}^2 + 2 \int_s^t e^{(2\alpha+1)(t-u)} (\mathbb{E}Y_u^k - \mathbb{E}Y_u^{n_k})^2 \, du,$$

which together with (14.10) and (14.11) implies that

$$\|Y_t^k - Y_t^{n_k}\|_{\mathrm{ms}}^2 \leq 4e^{(2\alpha+1)(t-s)}(|\alpha| + 2)^2 + \frac{2\delta^2}{2\alpha + 1} < \frac{\epsilon^2}{2} + \frac{\epsilon^2}{2} = \epsilon^2.$$

Here (14.9) was used to obtain the preceding inequality.

Thus the set $\{\phi(t, t_k, X_k)\}_{k \in \mathbb{N}}$ is covered by the finite union of open balls with radius ϵ centered at $\phi(t, t_k, X_k)$, where $k \in J(\lambda)$, i.e., it is totally bounded. It follows that the MS-RDS is asymptotically compact. Hence by Theorem 14.1 it has a mean-square attractor $\mathcal{A}$ with component subsets $A(t)$ for $t \in \mathbb{R}$ which contain the zero solution.

It remains to show that the sets $A(t)$ also contain points other than zero. Consider a uniformly bounded family of bounded sets defined by

$$D(t) := \left\{ X \in \mathfrak{X}_t : \mathbb{E}X = \sqrt{(\alpha+1)/2}, \ \mathbb{E}X^2 = \alpha + 1 \right\}.$$

Recall that these values are a steady state solution of the moment ODEs (14.3)–(14.4). Hence $\phi(t, s, X_s) \in D(t)$ for all $t > s$ when $X_s \in D(s)$. Then by pullback attraction

$$\mathrm{dist}_{\mathfrak{X}_t}(\phi(t, s, X_s), A(t)) \leq \mathrm{dist}_{\mathfrak{X}_t}(\phi(t, s, D(s)), A(t)) \to 0 \quad \text{as} \quad s \to -\infty.$$

The convergence here is mean-square convergence, and $\mathbb{E}\phi(t, s, X_s)^2 \equiv \alpha + 1$ for all $t > s$. Thus, there exists a random variable $X_t^* \in A(t) \cap D(t)$ for each $t \in \mathbb{R}$, i.e., the mean-square attractor is nontrivial.

It follows by the arguments in Chapter 4 that there is in fact a nonzero entire solution $\bar{X}_t \in A(t) \cap D(t)$ for all $t \in \mathbb{R}$. $\qquad \square$

Notes

1.   This book, like part of the monograph *Nonautonomous Dynamical Systems* by [Kloeden and Rasmussen (2011)], is based on lectures given in Frankfurt am Main by the first author and later in Wuhan by both authors. The first part of the book overlaps with the material in [Kloeden and Rasmussen (2011)], but also contains some new work, in particularly on an alternative construction of pullback attractors for strictly contractive processes and the construction of forward attractors.

2.   There are many books on dynamical systems, which usually means *autonomous* dynamical systems. [Hale (1988); Robinson (2001); Vishik (1992); Temam (1997)] are mainly concerned with attractors in autonomous systems.

A standard reference on ordinary differential equations with an orientation to dynamical systems is [Hirsch, Smale and Devaney (2004)]. [Aulbach (1998)] is highly recommended for those who can read German. [Kato, Martynyuk and Sheshakov (1996)] and [Lasalle (1976)] are mainly concerned with stability issues.

Properties of set-valued mappings are given in [Aubin and Franksowka (1990)].

3.   Applications of nonautonomous and random dynamical systems in the life sciences are given, e.g., in [Caraballo and Han (2016); Chesson (2017); Kloeden, and Pötzsche (2013a,b)]. See also the review article [Crauel and Kloeden (2015)]. Discrete time nonautonomous dynamical systems are treated in [Kloeden, Pötzsche and Rasmussen (2012)] and also [Pötzsche (2010)].

The SIR model is described in detail in [Brauer, van den Driessche and Wu (2008)]. Chapter 6 is based on the paper [Kloeden and Kozyakin (2011)]. See [Kloeden and Pötzsche (2015)] for bifurcations in the SIR models.

4.   [Carvalho, Langa and Robinson (2013); Chepyzhov and Vishik (2002)] deal with nonautonomous attractors of infinite dimensional nonautonomous dynamical systems, e.g., those generated by parabolic partial differential equations.

5.   The definition of process is due to [Dafermos (1971); Hale (1988)] and is the generalisation of the notion of semigroup corresponding to an autonomous system.

The definition of a pullback attractor was motivated by that of a random attractor in [Crauel and Flandoli (1994)]. In some earlier papers [Kloeden and Schmalfuß (1996); Kloeden and Schmalfuss (1997a,b); Kloeden and Stonier (1998)] they were called cocycle attractors. The name pullback attractor was later introduced in [Kloeden (2000)] to distinguish them from forward attractors. The existence of pullback attractors has been proved in many versions in the literature with the original proofs being based on those for the existence of random attractors, see the notes in [Kloeden and Rasmussen (2011)] for more information. Attraction universes and tempered bounded sets were introduced in [Schmalfuß (1997)] for random dynamical systems. See also [Kloeden and Rasmussen (2011)].

The existence of pullback attractors for uniformly contracting processes was first used in the random context in [Caraballo, Kloeden and Schmalfuß (2004)] and later in the deterministic setting by [Kloeden and Lorenz (2013)].

The analysis of the pullback attractor of the nonautonomous Bernoulli ODE (8.3) is taken from [Kloeden and Siegmund (2005)], where bifurcations in nonautonomous systems are explored.

The one-sided Lipschitz condition is well-known in numerical analysis. The one-sided dissipative Lipschitz condition is often called the uniform dissipativity condition.

Properties of set-valued mappings are given in [Aubin and Franksowka (1990)].

6. Skew product flows originated in ergodic theory and were extensively studied in connection with ordinary differential equations by [Sell (1967, 1971)]. Related work on limiting equations is presented in [Kato, Martynyuk and Sheshakov (1996)]. A class of ODEs with time variation more general than almost periodic is investigated in [Kloeden and Rodrigues (2011)]. These also generate skew product flows with a compact base space.

7. The material in the last three chapters is new and is based on papers in journals by the authors and their coworkers.

8. Chapter 10 on the limitations of pullback attractors is based on the paper [Kloeden, Pötzsche and Rasmussen (2012)]. See also [Cheban, Kloeden and Schmalfuß (2001)].

9. The construction of forward attractor was given by [Kloeden and Lorenz (2016)], see also [Kloeden, Lorenz and Yang (2016)]. Nonautonomous omega limit sets were discussed in [Crauel and Flandoli (1994); Lasalle (1976); Pötzsche (2010)]. They and a new concept of forward attracting set, which is a weaker version of the Haraux–Vishik uniform attractor [Chepyzhov and Vishik (2002); Haraux (1991); Vishik (1992)], were investigated in [Kloeden (2016); Kloeden and Yang (2016)]. Theorem 11.5 on asymptotically autonomous systems is a modification of a theorem in [Kloeden and Simsen (2015)]. Lemma 11.1 is taken from [Cui and Kloeden (2019)].

10. Asymptotic positive invariance for ODEs was introduced in [Lakshmikantham and Leela (1967)]. See also [Kloeden (1975)]. Related concepts *quasi-invariance* and *lifted invariance* for skew product flows were introduced by [Miller (1965)] and [Botolan, Carvalho and Langa (2014)], respectively.

Asymptotic positive and negative invariance for nonautonomous omega-limit sets were proposed in [Kloeden (2016)]. Properties of forward attracting sets and nonautonomous omega-limit sets such as asymptotical invariance and upper semi continuous dependence on a parameter were investigated in [Kloeden (2016)] and [Kloeden and Yang (2016)]. Corresponding results for a reaction diffusion equation on a time varying domain given in [Kloeden and Yang (2019)]. See also [Crauel, Kloeden and Yang (2011)].

11. The theory of random dynamical systems is expounded in [Arnold (1998)]. See also [Crauel and Flandoli (1994); Crauel and Kloeden (2015)]. There are many books on stochastic differential equations at different levels of abstraction, see for example [Arnold (1978); Kloeden and Platen (1992)]. Random ordinary differential equations are developed in [Han and Kloeden (2017)].

12. Mean-square random dynamical systems based on deterministic two-parameter semi-groups on a state space of random variables or random sets with the mean-square topology were introduced in [Kloeden and Lorenz (2012)]. Stochastic differential equations with nonlocal sample dependence, including expectation terms, were proposed in [Kloeden and Lorenz (2010)]. The example analysed in Chapter 14 comes from [Doan, Rasmussen and Kloeden (2015)].

Bibliography

L. Arnold *Stochastic differential equations: theory and applications* Wiley-Interscience, New York (1978).

L. Arnold, *Random Dynamical Systems*, Springer, New York (1998).

J. P. Aubin and H. Franksowka, *Set-Valued Analysis,*, Birkhäuser, Basel, 1990.

B. Aulbach, *Gewöhnliche Differentialgleichungen,* Spektrum der Wissenschaften, Heidelberg, 1998.

A. Babin and M. Vishik, *Attractors of Evolution Equations*, North-Holland, Amsterdam, 1992.

M. C. Bortolan, A. N. Carvalho and J. A. Langa, Structure of attractors for skew product semiflows, *J. Differential Equations,* **257** (2014), 490–522.

F. Brauer, P. van den Driessche and Jianhong Wu (eds.), *Mathematical epidemiology*, Lecture Notes in Mathematics, vol. 1945, Springer-Verlag, Berlin, 2008, Mathematical Biosciences Subseries.

T. Caraballo, P. E. Kloeden and B. Schmalfuß, Exponentially stable stationary solutions for stochastic evolution equations and their perturbation, *Applied Mathematics and Optimization,* **50** (2004), 183–207.

T. Caraballo and Xiaoying Han, *Applied Nonautonomous and Random Dynamical Systems,* Springer Briefs in Mathematics, Springer-Verlag, Heidelberg, 2016.

A. N. Carvalho, J. A. Langa and J. C. Robinson, *Attractors of Infinite Dimensional Nonautonomous Dynamical Systems,* Springer, New York, 2013.

D. N. Cheban, *Global Attractors of Non-Autonomous Dissipative Dynamical Systems*, volume 1 of *Interdisciplinary Mathematical Sciences*. World Scientific, Hackensack, New Jersey, 2004.

D. N. Cheban, P. E. Kloeden and B. Schmalfuß, Pullback attractors in dissipative nonautonomous differential *Journal of Dynamics and Differential Equations* **13** (2001), 185–213.

V. V. Chepyzhov and M. I. Vishik, *Attractors for Equations of Mathematical Physics*, Amer. Math. Soc., Providence, Rhode Island, 2002.

P. Chesson, AEDT: A new concept for ecological dynamics in the ever-changing world, *PLOS Biology*, 30 May 2017, doi.org/10.1371/journal.pbio.2002634.

I. Chueshov, *Monotone random systems theory and applications*, Lecture Notes in Mathematics, vol. 1779, Springer-Verlag, Berlin, 2002.

H. Crauel and F. Flandoli. Attractors for random dynamical systems. *Probability Theory and Related Fields*, **100** (1994), 365–393.

H. Crauel and P. E. Kloeden, Nonautonomous and Random Attractors, *Jahresbericht der Deutschen Mathematiker-Vereinigung*, **117** (2015), 173–206.

H. Crauel, P. E. Kloeden and Meihua Yang, Random attractors of stochastic reaction-diffusion equations on variable domains, *Stochastics & Dynamics* **11** (2011), 301–314.

Hongyong Cui and P. E. Kloeden, Comparison of attractors of asymptotically equivalent difference equations in *Difference Equations, Discrete Dynamical Systems and Applications*, S. Elaydi et al. (eds.), Springer Proceedings in Mathematics & Statistics 287, Springer Nature Switzerland (2019); pp. 31–50.

Hongyong Cui, P. E. Kloeden and Meihua Yang, Asymptotic invariance of omega limit sets of nonautonomous dynamical systems, *Discrete and Continuous Dynamical Systems, series S* **13** (2019), doi: 10.3934/dcdss.2020065.

C. M. Dafermos, *An invariance principle for compact processes*, Journal of Differential Equations, **9** (1971), 239–252.

S. T. Doan, M. Rasmussen and P. E. Kloeden, The mean-square dichotomy spectrum and a bifurcation to a mean-square attractor, *Discrete Conts. Dyn. Systems, Series B* **20**, (2015), 875–887.

J. K. Hale, *Asymptotic behavior of dissipative systems*, Amer. Math. Soc., Providence, 1988.

Xiaoying Han and P. E. Kloeden, *Random Ordinary Differential Equations and their Numerical Solution,* Springer Nature Singapore, 2017.

A. Haraux, *Systemes dynamiques dissipatifs et applications*, Research in Applied Mathematics, vol. 17. Masson, Paris, 1991.

M. W. Hirsch, S. Smale and R. L. Devaney, *Differential Equations, Dynamical Systems & an Introduction to Chaos,* second edition, Elsevier, Amsterdam, 2004.

J. Kato, A. A. Martynyuk and A. A. Sheshakov, *Stability of Motion of Nonautonomous Systems*, Gordon & Breach Publ., London, 1996.

P. E. Kloeden, Asymptotic invariance and limit sets of general control systems, *J. Differential Equations,* **19** (1975), 91–105.

P. E. Kloeden, Asymptotic invariance and the discretisation of nonautonomous forward attracting sets, *J. Comput. Dynamics,* **3** (2016), 179–189.

P. E. Kloeden, Pullback attractors in nonautonomous difference equations. *J. Difference Eqns. Applns.* **6** (2000), 33–52.

P. E. Kloeden and V. Kozyakin, The dynamics of epidemiological systems with nonautonomous and random coefficients, *MESA - Mathematics in Enegineering, Science and Aerospace* **2**, No. 2, (2011), 105–118.

P. E. Kloeden and T. Lorenz. Stochastic differential equations with nonlocal sample dependence. *Stochastic Analysis and Applications*, 28(6):937–945, 2010.

P. E. Kloeden and T. Lorenz. Mean-square random dynamical systems. *Journal of Differential Equations*, 253(5):1422–1438, 2012.

P. E. Kloeden and T. Lorenz, Pullback incremental attraction, *Nonautonomous & Random Dynamical Systems* (2013), 53–60. DOI: 10.2478/msds-2013-0004.

P. E. Kloeden and T. Lorenz, Construction of nonautonomous forward attractors, *Proc. Amer. Mat. Soc.,* **144** (2016), 259–268.

P. E. Kloeden, T. Lorenz and Meihua Yang, Forward attractors in discrete time nonautonomous dynamical systems in *Differential and Difference Equations with Applications*, Springer Proceedings in Mathematics & Statistics 164, Editors: O. Dosly, P. E, Kloeden, S. Pinelas; Springer, Heidelberg (2016), pp. 314–321.

P. E. Kloeden and E. Platen, *Numerical Solution of Stochastic Differential Equations* Springer-Verlag, Heidelberg, 1992; second revised printing 1999.

P. E. Kloeden and C. Pötzsche (Editors), *Nonautonomous Systems in the Life Sciences,*

Springer Lecture Notes in Mathematics (Biosciences series) vol. 2102, Springer-Verlag, Heidelberg, 2013.

P. E. Kloeden and C. Pötzsche, Nonautonomous dynamical systems in the life sciences, in *Nonautonomous Dynamical Systems in the Life Sciences*, (Editors: P. Kloeden & C. Pötzsche), Springer Lecture Notes in Mathematics (Biosciences series) vol. 2102, Springer-Verlag, Heidelberg (2013); pp. 3–38.

P. E. Kloeden and C. Pötzsche, Nonautonomous bifurcation scenarios in SIR models, *Math. Methods Appl. Sci.* **38** (2015), 3495–3518.

P. E. Kloeden, C. Pötzsche and M. Rasmussen, Discrete-time nonautonomous dynamical systems, in *Stability and Bifurcation Theory for Non-Autonomous Differential Equations*, (Editors: R. Johnson & M. P. Pera), Lecture Notes in Mathematics, vol. 2065, Springer-Verlag, Heidelberg, 2012, pp. 35–102.

P. E. Kloeden and M. Rasmussen, *Nonautonomous Dynamical Systems,* American Mathematical Society, Providence, 2011.

P. E. Kloeden and H. M. Rodrigues, Dynamics of a class of ODEs more general than almost periodic, *Nonlinear Analysis TMA*, **74** (2011), 2695–2719.

P. E. Kloeden and B. Schmalfuß, Lyapunov functions and attractors under variable time-step *Discrete and Continuous Dynamical Systems*, **2** (1996), 163–172.

P. E. Kloeden and B. Schmalfuß, Cocycle attractors of variable time-step discretizations of *Journal of Difference Equations and Applications*, **3** (1997), 125–145.

P. E. Kloeden and B. Schmalfuß, Nonautonomous systems, cocycle attractors and variable time-step discretization, *Numerical Algorithms*, **14** (1997), 141–152.

P. E. Kloeden and S. Siegmund, Bifurcation and continuous transition of attractors in autonomous and nonautonomous systems, *Inter. J. Bifurcation & Chaos* **15** (2005), 743–762.

P. E. Kloeden and J. Simsen, Attractors of asymptotically autonomous quasi-linear parabolic equation with spatially variable exponents, *J. Math. Anal. Appl.*, **425** (2015), 911–918.

P. E. Kloeden and D. J. Stonier, Cocycle attractors in nonautonomously perturbed differential equations. *Dynamics of Continuous, Discrete and Impulsive Systems*, **4** (1998), 211–226.

P. E. Kloeden and M. H. Yang, Forward attraction in nonautonomous difference equations, *J. Difference Eqns. Applns.*, **22** (2016), 513–525.

P. E. Kloeden and M. H. Yang, Forward attracting sets of reaction-diffusion equations on variable domains, submitted *Discrete and Continuous Dynamical Systems, Series B*, **24** (2019), 1259–1271.

O. Ladyzhenskaya, *Attractors for Semigroups and Evolution Equations*, Cambridge Univ. Press, Cambridge, 1991.

V. Lakshmikantham and S. Leela, Asymptotic self-invariant sets and conditional stability, in *Proc. Inter. Symp. Diff. Equations and Dynamical Systems*, Puerto Rico 1965, Academic Press, New York, (1967), 363–373.

J. P. Lasalle, *The Stability of Dynamical Systems*, SIAM-CBMS, Philadelphia, 1976.

R. K. Miller, Asymptotic behavior of solutions of nonlinear differential equations, *Transactions Amer. Math. Soc.* **115** (1965), 400–416.

C. Pötzsche. *Geometric Theory of Discrete Nonautonomous Dynamical Systems*, Lecture Notes in Mathematics, vol. 2002, Springer-Verlag, Heidelberg, 2010.

J. Robinson, *Infinite-Dimensional Dynamical Systems: An Introduction to dissipative Parabolic PDEs and the theory of Global Attractors*, Cambridge University Press, Cambridge, 2001.

B. Schmalfuß, The random attractor of the stochastic Lorenz system, *Zeitschrift für Angewandte Mathematik und Physik*, **48** (1997), 951–975.

G. R. Sell, Nonautonomous differential equations and dynamical systems. *Transactions of the American Mathematical Society*, 127:241–283, 1967.

G. R. Sell, *Topological Dynamics and Ordinary Differential Equations*. Van Nostrand Reinhold Mathematical Studies, London, 1971.

L. Stone, B. Shulgin, and Z. Agur, *Theoretical examination of the pulse vaccination policy in the SIR epidemic model*, Math. Computer Modelling **31** (2001), 207–215.

R. Temam, *Infinite-Dimensional Dynamical Systems in Mechanics and Physics*, Springer-Verlag, 1997.

M. I. Vishik, *Asymptotic Behaviour of Solutions of Evolutionary Equations*, Cambridge University Press, Cambridge, 1992.

Index